W9-CFX-648

HANDBOOK OF REMOTE CONTROL & AUTOMATION TECHNIQUES

No. 1077
$12.95

HANDBOOK OF REMOTE CONTROL & AUTOMATION TECHNIQUES

By John E. Cunningham

E 6 3

TAB BOOKS

BLUE RIDGE SUMMIT, PA. 17214

FIRST EDITION

FIRST PRINTING—DECEMBER 1978

Copyright © 1978 by TAB BOOKS

Printed in the United States of America

Library of Congress Cataloging in Publication Data

Cunningham, John, 1938-
   Handbook of remote control & automation techniques.

   Includes index.
   1. Automatic control. 2. Remote control.
I. Title.
TJ213.C86   620'.46   78-11475
ISBN 0-8306-9848-5
ISBN 0-8306-1077-4 pbk.

Cover artwork is courtesy of Digital Equipment Corporation. DIGITAL and VAX are trademarks of Digital Equipment Corporation.

Preface

It is amazing that so few electronics engineers, technicians, and hobbyists ever build electronic control systems to make life easier around their homes. Computer hobbyists in particular often tell of all the wonderful things their computers can do but few of these systems ever reach completion.

Perhaps this is due to the apparent reluctance of craftsmen to practice their crafts in their own homes. It is legendary that the pipes in a plumber's home often leak and that a carpenter's home is often in a state of disrepair. It may also be that a control system includes more than electronic components and the mechanical requirements are confusing to one whose expertise is primarily in the field of electronics.

There is no question that the addition of a few electronics systems would make life easier in any home. When family members are disabled in any way the systems are almost necessities. Although the types of automation and remote control systems used in government and industry are too complex and expensive for home use, there are many simple systems that can be built for the home at low cost. Several such systems are described in this book.

All of the systems described here have been built and tested. Most of them were built to make life easier for a girl who is confined to a wheelchair. In her case they were necessities for normal living. Hopefully most readers will find the systems conveniences rather than necessities. Another advantage of electronic control systems is that they enable us to demonstrate the wonders of electronics to friends who have little knowledge of the subject.

The first chapter of this book describes the basic principles of remote control and automatic systems. A functional analysis is given that relates the various parts of a control system and shows the similarity between different types of systems. Chapter 2 describes the types of sensing devices that can be used with most control systems.

The next three chapters deal with things that the average electronics technician is apt to be somewhat unfamiliar with, such as mechanical devices, hydraulic systems and electric motors. Emphasis is placed on the aspects of these devices that are involved in interfacing them with electronic systems.

Seven chapters are devoted to control devices of various types ranging from simple systems for controlling lights to systems that will permit opening doors by remote control.

Chapter 13 describes computer controlled systems and Chapter 14 deals with the problems of interfacing any of the systems in the book with microprocessor systems that are becoming very popular with hobbyists.

One of the problems facing the technician who attempts to build a system that uses more than simple electronic components is finding the mechanical parts necessary to complete the system and make it operational. Chapter 15 is devoted to this problem. Particular attention is given to adapting components and systems that were originally designed for other purposes to home control system.

The author would like to acknowledge the help and suggestions received from his associates, his students, and from his fellow radio amateurs. The greatest debt of gratitude is owed to Grace Slavik whose handicap served as the inspiration for many of these systems and whose stamina and cheerfulness in the face of adversity are an inspiration to all who know her.

Much credit must be given to the author's two "angels" Mrs. Janie Bowers and Mrs. Myrna Small who typed all of the manuscript and much of the rough draft. Without their help the book would have been impossible.

John E. Cunningham

Contents

1

Principles of Remote and Automatic Control Systems

Probably the best way to approach the subject of remote and automatic control systems is to start with a very simple, easy to understand, arrangement and then add features to make it more complex. All control systems have much in common and this simple example will make it easier to understand the more elaborate systems that follow.

One of the simplest control systems one can think of is an outside light which is controlled by a switch inside the house. The usual arrangment is shown in Fig. 1-1a. This is more properly called a remote switching system, rather than a remote control system, because the power line that operates the controlled device is carried right to the control point.

A true remote control system for an outside light is shown in Fig. 1-1b. Here the light is actually controlled by a relay which in turn is controlled by a low-voltage DC circuit. As simple as it is, this arrangement has many advantages over the remote switching circuit of Fig. 1-1a. For example, in as much as low voltage is used, light wiring can be employed without any risk of fire or electric shock. Furthermore many different switches can be added so that the outside light can be turned on or off from many different points inside the home.

The system of Fig. 1-1b can be made even more flexible by replacing the conventional relay with a latching relay that will latch

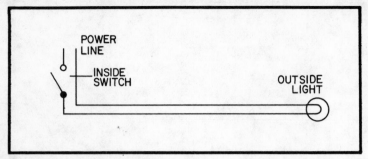

Fig. 1-1a. Simple remote switching system.

whenever it is actuated. With this setup there is no need to send a continuous control signal to keep the light on. All we have to do is to transmit a brief pulse that will latch the relay. To turn the light off we would transmit another brief pulse that would release the latch. We will discuss systems of this type in detail in a later chapter.

One limitation of the simple system just explained is obvious. Unless the switch is located where the outside light can be seen through a window there's no way of telling whether or not the light is on. What we need is some sort of indicator inside the house near the control switch that will tell us when the outside light is on. You will find that this is true of all but the most simple control systems. It is almost always necessary to have an indicator that will tell us the state of whatever it is that we're trying to control. In Fig. 1-1c we have added an indicator to our simple outside light control system. Here additional contacts on the relay drive a low voltage circuit that actuates a small indicator light inside the house. When the indicator light is on, the outside light is also on.

The arrangement of Fig. 1-1b or 1-1c is fully adequate for controlling an outside light but it does have limiatations. For example, there is no way of knowing if the outside light happens to be burned out. If the object being controlled was something much more important than a simple light, this could be a serious limitation. In such a case we may wish to use a more elaborate arrangement. A sensor such as a light detector might be used so we could be really sure that the outside light was on.

Without going into anything more complicated than a simple light that is controlled remotely we can see that it is possible to add many features to improve the system. For example, we might add a timer to the circuit and fix it so that the outside light would turn on for

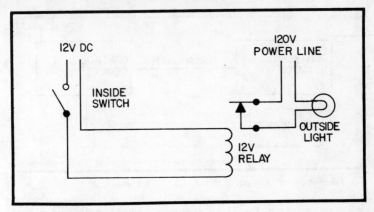

Fig. 1-1b. Low voltage control of outside lamp.

two or three minutes every time anyone rang the doorbell. This of course would mean that whenever the door bell rang the light would go on—night or day. If this were objectionable we could use a light sensor or another timer to be sure that the outside light only turned on after dark.

This example is admittedly trivial but it does serve to illustrate some of the functions that are common to all control systems. These

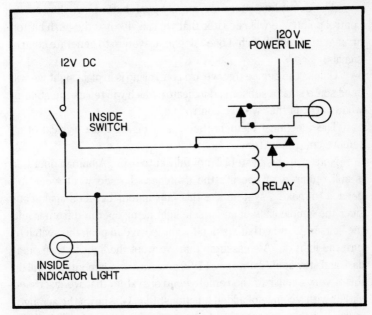

Fig. 1-1c. Remote control with indicator.

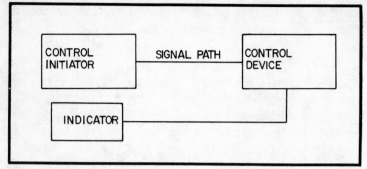

Fig. 1-2. Functional diagram of remote control system.

are summarized in the block diagram of Fig. 1-2. In any control system we must have some device that actually controls whatever it is we want controlled. In our example the control device is the relay that supplies power to a light. In most cases the control devices will be more complex. Connected to the control device we have a signal path which carries the control signal. The type of path used depends upon the nature of the control signal. To simplify the requirements for the signal path it is advisable to use low-voltage control signals. These may be either simple on-off DC signals or AC tones.

The control initiators are merely push button switches at the control point. We will see later that we can also use the push button arrangement of a Touch-Tone telephone system to generate control signals.

Unless the device that we are controlling is in plain sight we will need some sort of sensor and indicator which will tell us the state of whatever it is that we are controlling.

The functional elements shown in Fig. 1-2 are typical of all remote control systems.

Note that this system is not fully automatic. A human operator is still required to operate the switches. To see what would be needed to make a system like this fully automatic, let's again consider the simple case of an outside light being operated from inside the house. Assume that we don't want to have to push the switch to turn the light on. Also assume that we want the light to turn on at dark and stay lighted until say 11:00 pm. An arrangement that will do this is very similar to the remote control system that we discussed earlier with a couple of added functions. Fig. 1-3 shows the arrangement. A light sensor is located outside. This sensor will tell the system when it is dark outside. Instead of driving an indicator the output of the sensor is now used to generate a control signal that will turn on the outside light.

We can't use the sensor alone to turn on the outside light because we have also stipulated that for the light to turn on automatically it not only must be dark outside but must be before 11:00 pm. Therefore in addition to the signal from the light sensor we need another signal from a timer. The switches in the sensor and timer are connected in series as shown. When both devices are active, that is when it is dark outside and it is before 11:00 pm both on the switches will be closed and the light will be turned on.

CLOSED-LOOP AND OPEN-LOOP SYSTEMS

Fully automatic systems, that is systems that can operate without the intervention of a human operator can be classified into two general categories. These are closed-loop and open-loop systems. Fig. 1-4 shows a block diagram of a closed-loop system. The example we have chosen is the familiar household thermostat that is used with a furnace to control the temperature in a home.

The sensing element of the system is a thermostat. This device senses the temperature in the house and compares it with the desired temperature that is set by means of a dial. The thermostat generates a signal that corresponds to the difference between the actual temperature in the house and the desired temperature set on the dial. This signal is usually known as an *error signal*.

One thing that the error signal must do is tell the furnace the direction of the error; that is, it must tell whether the temperature in

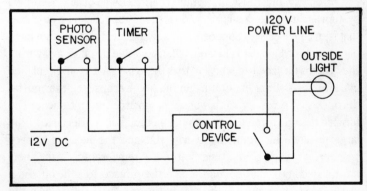

Fig. 1-3. Automatic control of outside lamp.

the house is higher or lower than the desired temperature. Otherwise the furnace would not know whether it should turn on or off. The error signal in some systems may also tell how great the error is. That is, the signal may be proportional to the amount that the actual temperature differs from the desired temperature.

The next element of the system is the transmission channel by which the error or control signal is transmitted to the device that actually controls the furnace. In the case of the ordinary household thermostat, the error signal is usually a low-voltage AC signal. The channel is a pair of small wires.

The final element of the control system is the relay that actually turns the furnace on or off. This is the output or control device of the system.

What we have shown in Fig. 1-4 is a very elementary temperature control system. In most household temperature control systems many other features are included such as an arrangement that will allow the forced air fan on the furnace to run after the burner has been turned off. We need not concern ourselves with these other features at this time; our simple system will serve to demonstrate the principle of the closed-loop system.

This type of system is called a "closed-loop" system because there is a continuous path around the entire control system. The input to the system is the temperature in the house which is converted into an electrical signal that controls the furnace. The furnace generates heat which, in turn, controls the temperature in the room. And as the temperature is the input to the system there is a complete closed loop around the whole system.

Probably the most important characteristic of a closed-loop system as far as the designer is concerned is that it can be unstable; that is, it can oscillate. Suppose, for example, that the thermostat in our system of Fig. 1-4 is located a considerable distance away from the nearest furnace register and that there isn't much air circulation in the area. As the temperature in the home drops, the thermostat developes an error signal that asks the furnace for more heat. The furnace then turns on and begins heating the house. Because of the large distance between the thermostat and the nearest furnace register, it will be a long time before the temperature at the thermostat becomes high enough to tell the furnace to shut off. As a result, the temperature in the rest of the area, particularly near the

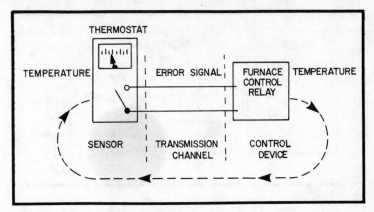

Fig. 1-4. Closed loop control system.

registers, will become much higher than the desired temperature. Now the house begins to cool. Let's say that any heat leaks such as doors or windows are closer to the registers than they are to the thermostat. The result is that by the time the temperature at the thermostat becomes low enough for it to ask the furnace for more heat, the temperature in the rest of the house will be well below the desired temperature. The result is that instead of the automatic control system providing a nice even temperature, it will continually oscillate between the two conditions—too hot and too cold.

In a temperature regulating system, such as the one we have described, instability is annoying but not particularly dangerous. But if the system were controlling the position of a door and the instability meant that the door continually swung open and closed, the system would not only be worthless, but positively hazardous.

Almost all closed-loop systems have the potential of instability. In the design of large control systems for industrial equipment one of the most important considerations is making sure that the system will be stable under all conditions. This usually isn't a major problem in most home or office control systems because there are usually comparatively simple ways of assuring that the system will be stable. Often this consists of opening the control loop giving us an open-loop system.

Figure 1-5 shows an example of an open-loop control system. The system is automatic in that it will operate automatically without any human intervention. The purpose of the system is to turn on a coffee maker at 6:00 am every morning. The thing that is sensed is

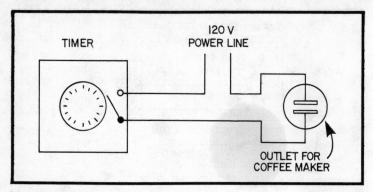

Fig. 1-5. Open-loop control system.

not the condition of the coffee pot, but simply the time of day. This is sensed by a timer which is actually an electric clock. Control is initiated by a pair of contacts that can be set to close at any desired time. At 6:00 am every morning, the contacts close turning on the coffee maker. If we wish, we may have another pair of contacts to turn it off at some later hour. This is an open-loop system because there is no closed path around the system. The input to the system is the time of day. The output is turning on a coffee maker. The condition of the coffee maker can in no way control the time of day so there is no closed path around the system.

The open-loop system is always stable. It can't break into undesirable oscillations. Its biggest drawback is that in many cases it has no way of knowing the present condition of whatever it is that we wish to control. This isn't significant in the case of the coffee maker because someone must fill the coffee maker the night before and be sure that the switches are in the proper positions and it is ready to go. It could be significant in a system that closed a door at the same time every day, because it might try to close a door that is already closed causing some damage.

In many applications an open loop system is entirely adequate. In other cases a closed-loop system is much better suited to the application. The decision of which type of system best suits a particular control problem will become clearer in later chapters.

PLANNING A CONTROL SYSTEM

The best way to tackle the design of a remote or automatic control system is to start with the output or control device. This is the device that we use to operate whatever it is that we want to control. This is where most electronics technicians or hobbyists get into trouble. Very few of them have had much practical experience outside of electronic circuits. Usually once the control device has been selected, the rest of the design is comparatively simple.

The easiest case to handle is where the device that we wish to control is electrically operated. Here we can use an electrical device such as a relay to accomplish the control function. The design of such a system is usually quite straightforward. The trouble usually arises when we are trying to control the physical position of something ranging from a door or window to draperies. Here we need a control device that will not only provide a mechanical motion, but the right kind of motion in the right direction. This field is usually foreign territory to the average electronics technician. For this reason we have devoted a great deal of attention to this phase of the problem in the following chapters.

The way to start solving the problem is to select a mechanical arrangement that will produce the desired motion. The device must be able to be powered with a readily available source of mechanical power such as an electric motor or solenoid.

Once the control device is selected, a method of sending control signals to it must be provided. This is usually simply a matter of finding a way to use a low-voltage AC or DC signal to control a larger voltage. The different ways that this can be done are described in detail later on. The signalling system should always use low voltage and current so that the wiring will not be critical and will not constitute a fire or electric shock hazard.

In most applications the thing that is being controlled will not be visible from the control point so it will be necessary to provide a sensing and indicating system. This is how the operator of a remote control system will know what he is doing and it is the feedback portion of a fully automatic control system.

The next decision that must be made is just how elaborate the system will be. Will it be a remote control system where the various devices are controlled from a single control point? Will there be several control points? Will some of the devices be controlled automatically to operate at a certain time of day? Or will it be an elaborate fully automatic system where everything that is controlled is related to something else? It is these considerations that dictate the complexity of the system.

DECIDING WHAT TO CONTROL

One of the most confusing aspects of designing a control system is deciding just what things should be controlled by the system. After all, you have probably been living for many years without any control systems around the home. A home control system is one of those things like an automatic transmission in an automobile that you never miss if you have never had it. It is only after living with a control system for a while that you begin to wonder how you ever managed to get along without it.

Once most technicians become interested in a home control system there is a tendency to want to control or automate almost everything in sight. The result is a plan for a system that is so elaborate that its chances of every being completed to the point where it is operational are negligible. It is better to start with a simple system that can be expanded later.

One of the best ways to decide what should be controlled is to carefully define your reasons for wanting a control system in the first place. If the reason is a physical handicap, the decisions are usually obvious. There are many things around a home that a handicapped person simply cannot cope with without some aid or assistance. If a handicapped person cannot move about without considerable difficulty, the system should control as many things as possible from one or more central locations. These things can range from turning lights on and off and automatically unlocking or opening doors to controlling the volume of a stereo or a TV set.

When the purpose of the system is personal convenience, the selection of just what functions should be controlled is much more difficult. It is really after you have the convenience of controlling many things from an easy chair that you begin to realize the value of such a system. The coffee lover, like the writer, who has once had a system that has the coffee ready before the alarm clock sounds will have great difficulty in going back to the old way of life where coffee had to be made after arising. One way of tackling the problem is to make a list of inconveniences. These can range from preparing the coffee in the morning and answering the door, at inconvenient times, to getting up at night to let the cat in. A little conscious attention to the inconveniences of everyday life will soon result in a list of things that might well be handled by a control system.

One of the strongest motives behind all sorts of human behavior, but one that is rarely acknowledged openly, is the desire to favorably impress other people. The motive is rarely acknowledged because it seems childish to go to a lot of trouble to build a system to simply favorably impress others with your knowledge and expertise. Although the motive is rarely acknowledged, much of our behavior ranging from dress style to table manners is geared primarily to group acceptance.

If impressing others with your expertise is one of your motives you don't have to acknowledge it openly. Neither is there any reason to be ashamed of it. After all, you have spent a great deal of time and effort to reach your state of electronic competence. If your accomplishments are confined to knowledge, you will have great difficulty sharing them with nontechnical people. If, on the other hand, your knowledge and ability result in a system that will make life more convenient, you can indeed show it to nontechnical people with justifiable pride.

One electronics engineer with a great deal of ingenuity won his wife with the aid of an automatic control system. The system was initiated by a timer that started the functions on the hour—when TV programs change. The system turned on the stereo and gradually increased its volume, descreased the volume of the TV audio, and dimmed the lights to a more cozy level. After a while the TV turned off. All of this was accomplished without our hero leaving his cozy position on the sofa.

An automatic control system is a natural for the computer hobbyist. The average computer buff will talk for hours about all the wonderful things that his computer can do for him around the house. Only too often, all it will really do is play games and display fancy graphics on a screen. These are not particularly impressive to other computer hobbyists who have similar systems and all too often they fail completely to impress anyone who has little technical knowledge. A nontechnical visitor is usually much more impressed when a computer does something he understands, like turning on a light, than he is by some data transformation that is completely incomprehensible to him.

Table 1-1 lists many of the things around the home that lend themselves to automatic or remote control. From this list you can probably recognize those things that are worth the trouble of controlling electronically.

Table 1-1. Functions Suitable for Control in a Home

OUTSIDE LIGHTS	STEREO ON-OFF
INSIDE LIGHTS	STEREO VOLUME
FANS	GARAGE DOOR
AIR CONDITIONER	OTHER DOORS
TV ON-OFF	DRAPERIES
TV VOLUME	COFFEE POT
INTERCOM SYSTEM	WINDOWS
DOOR LOCKS	

If a system is properly designed, it will be sufficiently flexible that functions can be added as time passes. There is no reason why the whole thing has to be operational at the start. Perhaps the system will only handle a single light when it is first installed. If enough control positions are provided on the control box, and if a sufficient number of wires are run when it is installed, it can be expanded as time permits to control more and more things.

Probably the best advice that can be given to the technician or engineer planning a system for the home is contained in the so-called KISS formula. KISS stands for "Keep It Simple, Stupid." The biggest reason for many proposed systems never reaching completion is that in the planning stage they become so elaborate that they would require many man years of effort to build and get working. The simpler a system is, the better chance it has of being completed.

Simplicity has many other advantages. The simpler a system, the easier it will be to troubleshoot it in the event of failure. Furthermore, a simple system with few components will be much more reliable.

2

Sensing and Indicating Devices

We pointed out earlier that for most remote control systems and all automation systems we must have a device that will sense the status of whatever it is we are controlling. With remote control systems we must also have an indicator that will show us this status. If we are to remotely turn on a light, or open a door, we must first know whether the light is on or off, or whether the door is open or closed.

The function of the sensor is to sense the status of what we are controlling and provide an electrical signal that is a function of the status. The easiest sensors to build are completely electrical ones. These are used when the quantity that we wish to sense is an electrical quantity. In this case all we have to do is convert from one electrical quantity to another, usually from one voltage level to another.

When the quantity we wish to sense is not electrical, the problem is more complicated. To take a simple example, suppose that we are controlling a door with a device that will automatically open and close it. We need a sensor that will respond to the position of the door and provide an electrical signal that we can use to determine the position of the door. Converting a mechanical quantity to an electrical quantity is usually much more involved than simply changing a voltage level.

SENSOR SPECIFICATIONS

It is very important that we know just what information we need from a sensor. If we can get by with a simple on-off indication the problem will be simplified considerably. On the other hand, if we need a continuous indication that is quite accurate designing and building the sensor can be quite a task.

In any case we usually want the signal from the sensor to be a low-voltage DC or AC signal that can be handled without any special wiring or insulation.

ELECTRICAL SENSORS

An electrical sensor is used whenever what we want to sense is an electrical quantity. We would use such a sensor whenever we want to know whether or not electrical power is being applied to the device we are controlling. Devices of this type include lights, fans, air conditioners, pumps and other electrically operated devices.

A typical example might be where a push button in the living room is used to control an air conditioner or fan in a bedroom in another part of the house. You would probably use the control system to turn on the air conditioner say half an hour or so before bedtime on hot summer nights. The sensor and indicator are needed so you will know whether or not the air conditioner is turned on. With several people living in a house anyone may think that someone else has turned on the air conditioner. The indicator will show for certain whether the air conditioner is on or off.

Figure 2-1 shows a simple arrangement that can be used to sense whether or not power is being applied to an appliance. At (a) in the figure the primary of a step-down transfomer is connected across the power line feeding the appliance we are controlling. The low voltage from the secondary is fed back to the control point to show us when the appliance is energized. All that this voltage will be used for is to operate a small indicator lamp, so a very small transformer can be used. The only reason for using the transformer in the first place is to isolate the indicator from the power line so that we can use a low signalling voltage.

In Fig. 2-1b we have added a rectifier to our sensing transformer. This simply converts our sensing signal from AC to DC. About the only time we would need to do this would be if our signal was to operate a DC relay. We might also wish to use DC indicating

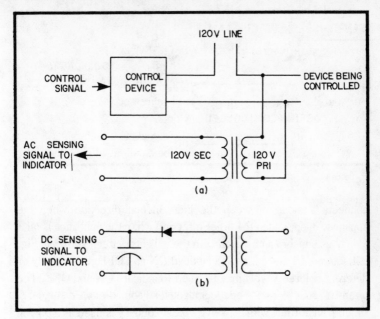

Fig. 2-1. Simple power sensing circuits.

signals if the signal wires were run in the same cable with intercom wires that might otherwise pick up hum.

Figure 2-2 shows another power sensing system. Here an AC relay is connected across the power line feeding the appliance that is remotely controlled. A relay has some advantages over the transformer of Fig. 2-1. With a relay as a sensor, we can use any type of sensing signal that we wish, and it is easy to get both an on and an off indication.

Referring to Fig. 2-2, we can run three wires from the sensing relay to the control point. When the circuit is closed between lines A and B, it means that the appliance is on. When the circuit is closed between lines B and C, it means that the appliance is off.

In applications where the device being controlled is a long distance from the control point, running three wires as in Fig. 2-2 can be expensive. By adding four diodes to the circuit we can do the same thing with only two wires. Figure 2-3 shows the arrangement; here our sensing relay is used to connect one of two diodes into the circuit. When the armature of the relay is down, diode D1 is connected between the two sensing lines. When the armature is up, diode D2 is connected between the two sensing lines. This in effect

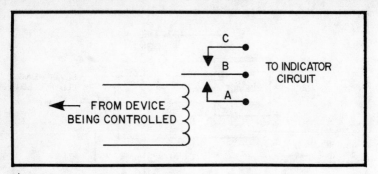

Fig. 2-2. Relay used as a power sensor.

connects a diode between the lines in one direction when the appliance is on, and in the other direction when the appliance is off.

This sensor can be used with the indicator circuit shown in Fig. 2-4. Lamps L1 and L2 can be labelled ON and OFF respectively or different colored jewels may be used to indicate ON and OFF. The on and off conditions can be recognized for a considerable distance. A single transformer with a twelve volt secondary is used as a power supply. The coil of relay K1 is connected across the appliance we are controlling. Consider first the situation where the controlled device is on and diode D1 is connected in the circuit. During the half cycle when the ungrounded side of the transformer secondary, point A in the figure, is negative with respect to ground no current will flow through either of the lamps because diode D1 will be reverse biased. Now during the opposite half cycle when point A is positive with respect to ground there will be a complete circuit through diodes D1 and D3 and through lamp L1.

When the controlled device is off and diode D2 is connected in the circuit by the relay the situation is somewhat similar. When point A is positive with respect to ground there will be no complete circuit. When point A is negative with respect to ground there will be a complete circuit through lamp L2 and diodes D4 and D2.

The diodes used in this circuit are the garden-variety silicon power supply diodes. About the only specification of interest is the reverse bias voltage rating. This should be 25V or greater. These diodes are very inexpensive and the features of the circuit make it worthwhile.

In the figure, we have specified 12-volt supplies and lamps. There is no particular reason why we have to adhere to this voltage.

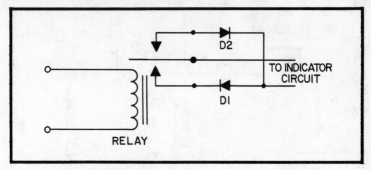

Fig. 2-3. Relay and diode used as a sensor.

Any voltage between about 6V and 24V can be used. The lamps must be rated for use at the power supply voltage.

Another good reason for using relays as sensing devices is that it is not necessary to have a voltage on the sensing lines all of the time. The signal, if we can call it that, from the relay is basically a contact closure. No power supply is used at the sensor, only at the indicator at the control point. It is possible to use a single indicating arrangement to tell us the status of several different devices. Figure 2-5 shows a circuit for this purpose. Here there are several different sets of input lines. Each of these comes from a sensor circuit like that shown in Fig. 2-3. These sensor circuits are connected, one at a time, to an indicator circuit like that used in Fig. 2-4. As the selector

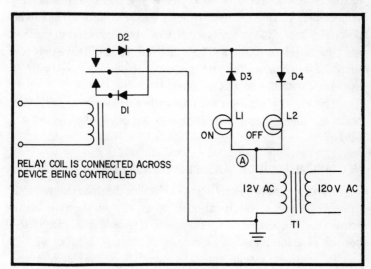

Fig. 2-4. Sensing circuit of Fig. 2-3 with AC power supply and indicating lamps.

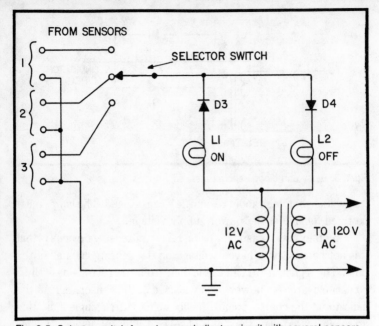

Fig. 2-5. Selector switch for using one indicator circuit with several sensors.

switch is turned to the different positions, either the ON lamp or the OFF lamp will light showing the status of the particular sensor.

This arrangement makes it possible to use a very small control box to control a large number of different devices. The indicating arrangement can also be used with one of the control circuits to be described later. The control circuitry can be connected to another gang on the selector switch. Thus the control box will have only two lamps, a selector switch and a couple of push button switches associated with the control circuitry.

The sensing arrangements described above will handle just about any sensing problem where the quantity to be sensed is a voltage.

MECHANICAL POSITION SENSORS AND INDICATORS

There are many control situations where the thing that we wish to sense is not an electrical quantity at all, but the position of something. For example, we may need a remote indication of the position of a door, window or drapery.

In many applications the mechanical sensing problem can be simplified considerably because we do not need a continuous indication of position, but only whether or not an object is in a particular position. For example, all that we might need to know is whether a door is fully closed or not. If it is open, we may not need to know just how far it is open. Or maybe we only have to know the position of something in a sort of rough manner. If this is the case, we can use switches as sensors. This is much simpler than using a sensing device that will give a continuous indication of the position of something. There are many commercially available switches that can be used for this purpose or you can build your own.

Some of the switches that are commonly used for position sensing are shown in Fig. 2-6. Figure 2-6a shows a magnetic door switch. The switch comes in two parts. One part containing the switch mechanism is mounted on the door frame. The other part containing a small permanent magnet is mounted on the door itself.

The actual switch mechanism is a magnetically operated reed. When there is a magnetic field in the vicinity of the reed, the switch will close. When the magnetic field is removed, the switch will open. The switch is mounted in such a way that when the door is closed, the magnet will be near the switch causing it to close. When the door is opened, the magnet will move away causing the switch to open. These switches are available in either normally open or normally closed configurations. You can choose the configuration best suited to a particular application.

Figure 2-6b shows what is usually called a snap-action switch. The switch is equipped with a small lever that touches the object that we wish to sense. When the lever is moved, the switch is actuated. Like the magnetic switch, the snap-action switch is available in either a normally-open or normally-closed configuration. Snap-action switches are also available in more complex configurations such as SPDT and DPST.

Snap-action switches can be obtained with many different lever arrangements so that they can be adapted to different applications. As a rule they require only a very small force or a very small displacement or both to actuate them. Several different snap-action switches can be placed so that they will indicate several different positions of an object.

Figure 2-6c shows a mercury switch. This switch consists of two electrodes, which do not touch each other, and a small pool of mercury. When the switch is positioned so that the mercury con-

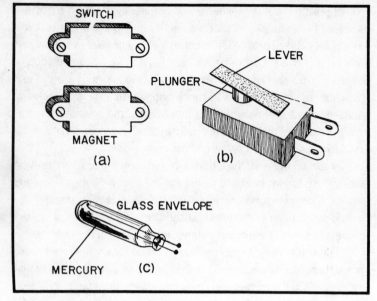

Fig. 2-6. Position sensing switches.

tacts both electrodes, the circuit is closed. Mercury switches are very handy for indicating when something is tilted.

CONTINUOUS POSITION SENSORS

The switches described above will usually indicate only one position of an object. For example, a magnetic door switch will indicate that a door is either closed or is not closed. If the door is open, it won't tell us how far open. A switch such as this is adequate for many control situations.

There are some control applications where we need to know the exact position of some object. Again, a door is a good example. The most obvious sensor to use for such an application is a potentiometer. To use a potentiometer for an application like this we need a mechanical arrangement that will translate the motion of the door into a rotary motion.

Figure 2-7 shows a potentiometer with a pulley on its shaft. The pulley is turned by a cord which is attached to the door. A weight on the free end of the cord keeps the cord tight. When the system is properly set up, the arm of the potentiometer will be at one extreme of its range when the door is fully opened and at the other extreme when the door is fully closed.

With a source of voltage connected across the potentiometer, the voltage appearing at the arm will be a function of the position of the door. If a linear potentiometer is used, we can connect a meter as shown to indicate the position of the door. If you wish, the scale of the meter can be marked to show the door position.

The potentiometer sensor can, of course, be used with objects other than doors. It could, for example, be used to indicate the position of a damper in a ventilating duct.

Theoretically, the potentiometer is an excellent sensor. In practice, it has many limitations. The mechanical linkage is usually the problem. Unless great care is taken in design and installation the cord may slip or break and the arrangement may require a great deal of attention. If we don't need to know the exact position of an object, but can get by with an approximate indication we can use a much simpler arrangement.

The mechanical portion of the sensor shown in Fig. 2-8 is somewhat similar to that of Fig. 2-7. Here again we use a cord to get a linear mechanical motion that is a function of the position of a door.

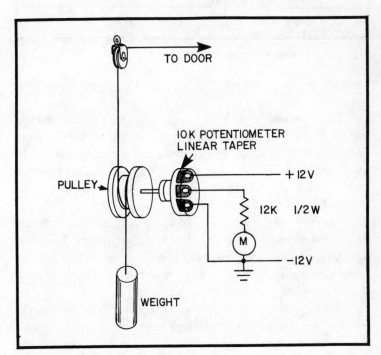

Fig. 2-7. Position sensing potentiometer.

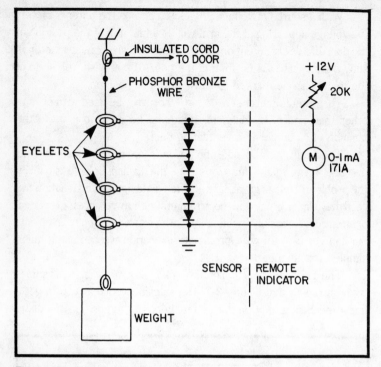

Fig. 2-8. Diode switching position sensor.

Here the similarity ceases. We don't need a potentiometer or a pulley.

The arrangement of Fig. 2-8 operates on the principle that the voltage drop across a forward-biased diode is reasonably constant. With a silicon diode of the type used as a rectifier in a power supply this voltage is about 0.7V. In our circuit, the cord which is fastened to the door is tied to a flexible wire. Phosphor bronze wire of the type used to hang pictures is fine for the purpose.

The cord and the wire pass through several eyelets. Between each pair of eyelets we have connected two silicon diodes. The diode string is connected through a dropping resistor to a voltage source. A line from the top of the diode string is connected to a voltmeter at the control point of the system as shown in Fig. 2-8.

When the door is closed, the wire that is tied to the cord shorts out all of the diodes. Thus the voltage across the string is zero and the voltmeter will indicate zero. When the door is partially opened, the cord, which is an insulator, passes between two of the eyelets.

32

The two diodes connected between these eyelets are no longer shorted and the voltage across the diode string will be equal to two diode-drops—about 1.4V.

As the door is opened more, more of the diodes will be switched into the circuit. When the door is fully opened the voltage across the string of diodes will be equal to eight diode-drops—about 5.6V. If you adjust the variable resistor so that you get full scale indication of the voltmeter when the door is fully opened, the deflection of the voltmeter will be proportional to the amount that the door is opened.

In our example, we used four sets of diodes. We used two diodes in each position rather than one simply to increase the signal to the voltmeter. We could calibrate the scale of the voltmeter to read "CLOSED—¼—½—¾—OPEN."

There is no reason why the system should be restricted to four steps. If we need more steps, we can simply add more diodes and more eyelets.

When the object whose position we are sensing tilts, we can use the same arrangement with mercury switches across the diodes rather than eyelets. A typical application of this type involves sensing the position of an overhead garage door. When a door like this opens, first the top panel, then the others, tilt from a vertical to a horizontal position.

You can use mercury switches with this kind of a door by means of the arrangement shown in Fig. 2-9. Here a mercury switch is connected across one or two diodes and the switch is connected to an ordinary terminal strip. The terminal strip is attached to the panel

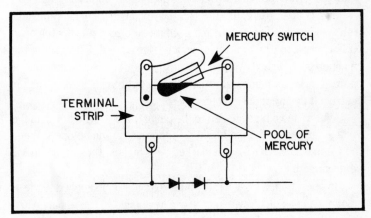

Fig. 2-9. Mercury switch tilt sensor.

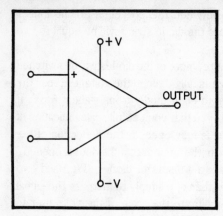

Fig. 2-10. Voltage comparator.

of the door and the angle at which the switch is mounted is set so that the switch will open when the door panel starts to tilt.

You can use as many of the arrangements of Fig. 2-9 that you wish to get as much resolution of the door position that you need for a particular system.

Naturally, these switching arrangements are not restricted to sensing door positions. They can be used wherever the motion of the object you are sensing can be translated into a tilting motion.

THE VOLTAGE COMPARATOR

A voltage comparator is an integrated circuit that is used to compare the magnitudes of two voltages. It can be used with light emitting diodes (LED) to provide a graphic indication of the position of anything that can open and close switches. As shown in Fig. 2-10, a voltage comparator consists essentially of a very high gain amplifier with two inputs. The output pin is connected to the collector of an NPN transistor, the emitter of which is grounded. The gain of the amplifier is so high that the output transistor is either saturated or open. Thus the comparator can be thought of a switch that is controlled by the input voltages.

When the voltage applied to the pin marked + is higher than the voltage applied to the pin marked − the output will be open. When the voltage applied to the pin marked − is higher, the output pin will be grounded.

In operation, a known reference voltage can be applied to the + input pin and an unknown voltage can be applied to the − pin. When the unknown voltage is lower than the reference voltage, the output pin will be open. When the unknown voltage is higher than the reference voltage, the output pin will be grounded.

The type 339 integrated circuit contains four separate voltage comparators and is called a quad comparator. Figure 2-11 shows the pin connections and maximum ratings of this circuit. The price is

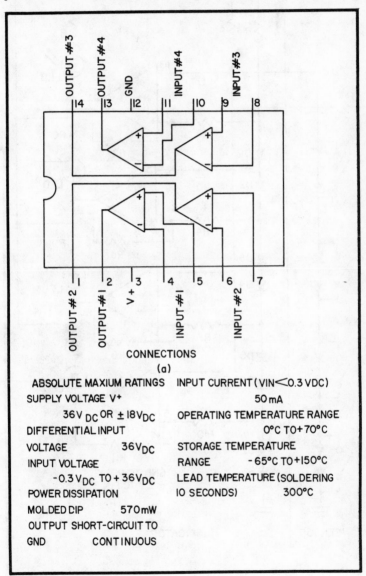

CONNECTIONS

(a)

ABSOLUTE MAXIUM RATINGS

SUPPLY VOLTAGE V+

    36V $_{DC}$ OR ± 18V$_{DC}$

DIFFERENTIAL INPUT

VOLTAGE   36V$_{DC}$

INPUT VOLTAGE

    -0.3 V$_{DC}$ TO + 36V$_{DC}$

POWER DISSIPATION

MOLDED DIP   570mW

OUTPUT SHORT-CIRCUIT TO

GND   CONTINUOUS

INPUT CURRENT (VIN < 0.3 VDC)

    50 mA

OPERATING TEMPERATURE RANGE

    0°C TO +70°C

STORAGE TEMPERATURE

RANGE   - 65°C TO +150°C

LEAD TEMPERATURE (SOLDERING

IO SECONDS)   300°C

Fig. 2-11. Type 339 quad comparator.

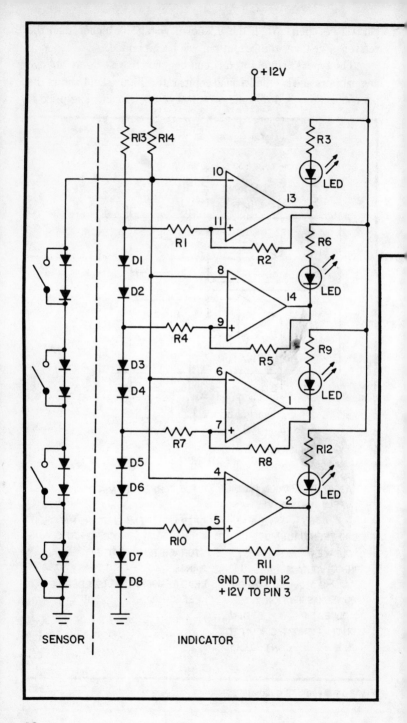

+12V

R13 R14   R3   LED

10
-
11   13
+
R1
R2   R6   LED

DI
D2
8
-
14
9
+
R4
R5   R9   LED

D3
D4
6
-
1
7
+
R7
R8

D5   R12   LED
D6
4
-
5   2
+
R10
R11

D7   GND TO PIN 12
D8   +12V TO PIN 3

SENSOR   INDICATOR

36

PARTS LIST

DI-D8	SILICON DIODES**
LEDS	LIGHT EMITTING DIODES, TYPE NSL 102*
RI, R4, R7, RII	100, 1/2 WATT RESISTOR
R2, R5, R8, RII	100K, 1/2 WATT RESISTOR
R3, R6, R9, RI2	470, 1/2 WATT RESISTOR
RI3	10K, 1/2 WATT RESISTOR
RI4	12K, 1/2 WATT RESISTOR
COMPARATOR	TYPE 339 COMPARATOR, RADIO SHACK 276-1712

*BULK PACKAGES OF LEDS ARE AVAILABLE FROM RADIO SHACK

** ANY SILICON DIODE HAVING AN INVERSE VOLTAGE RATING GREATER THAN 25V CAN BE USED. FOR EXAMPLE RADIO SHACK. PART NO 276-1122 COVERS A PACKAGE OF SEVERAL SUITABLE DIODES.

Fig. 2-12. Graphic indicator using LEDs.

quite low and the device is well suited to drive LEDs in a position indicator.

GRAPHIC POSITION INDICATOR

Figure 2-12 shows the diagram of a graphic position indicator using the Type 339 quad comparator. The sensor is the same as that used in Fig. 2-9 with one additional pair of diodes. A voltage that is a measure of the position of something such as a garage door is generated by switching diodes in and out of a circuit. The circuit of Fig. 2-12 uses the four voltage comparators in a quad comparator such as the Type 339, The reference voltages are generated from a string of diodes that is nearly identical to the diode string used in the sensor. These reference voltages are applied to the + inputs of the comparators. The – inputs are connected in parallel to the voltage generated by the sensor.

To understand how the circuit works let's start with the situation where all of the switches in the sensor are closed. Under this condition, all of the comparator – inputs are less than the reference voltages. Thus the outputs of the comparators will be high. That is, the output pins will be open circuits and no current will flow through the LEDs. Now let's consider the situation when the bottom switch in the sensor portion of the circuit is open. Now the voltage from the sensor will exceed the reference voltage applied to the bottom comparator in the figure. The presence of the extra diode at the bottom of the sensor circuit will assure this. The output of the bottom comparator will now be grounded and current will flow through the LED that is connected to it. Now when the next switch in the sensor opens a similar situation will prevail at the second comparator from the bottom and its LED will light. Thus the string of LEDs will light up starting at the bottom of the figure as the corresponding switches in the sensor are opened.

The LEDs can be arranged on a panel to give a graphic display of the position of the object that is connected to the sensor. For example, if the sensor switches are mercury switches attached to the panels of an overhead garage door, the diodes can be arranged in a vertical line as shown in Fig. 2-13a. Here, when the door is fully closed, all of the LEDs will be out. When the door is about one quarter open, the bottom LED will light. When the door is half open, the bottom two LEDs will light. When the door is three quarters open, the bottom three LEDs will light and when the door is fully open all of the LEDs will light. Thus the display will graphically resemble the position of the door. If the sensor is used with an ordinary door, the LEDs can be arranged in an arc as shown in Fig. 2-13b to provide a graphic representation of the position of the door.

In Fig. 2-12 we used four comparators and four LEDs. This is probably enough for almost any position indication required in a home control system. It can be expanded, however, by adding four more comparators, giving eight increments of indication. In Fig. 2-8, we used two silicon diodes for each switch position. This gives us voltage increments of about 1.4 volts. This really isn't necessary. We could use one diode across each sensing switch giving a voltage of about 0.7 volt. The only reason for using two diodes across each switch with the correspondingly greater voltage increment is that the circuit is easier to adjust and is less susceptible to noise.

The circuit of Fig. 2-12 isn't very critical. The two resistors used at the + input of each comparator provide a little positive feedback that will make the comparator switch very rapidly from the on to the off condition and vice versa. This will keep the comparators from oscillating even though the gain of the amplifiers is very high. The diodes are not at all critical. About the only requirement is that they all be silicon diodes to provide the constant voltage drop of 0.7V and that the inverse voltage rating be about 25 volts or greater. The

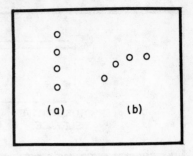

Fig. 2-13. LED arrangements for circuit of 2-12.

LEDs are not critical either. Just about any LED can be used in the circuit. If the current rating of the LED is very small, the resistors in series with them can be increased in value. Higher current LEDs can also be used by decreasing the value of the series resistors. Each comparator will ground a current of about 15 mA.

LIGHT SENSORS

One of the things that you might want to sense in a control system is light. Many control systems will operate in one mode during daylight hours and in another mode after dark. For example, there is no need for a system to automatically turn on outside lights if it is daylight anyway. On the other hand it might be desirable for the system to turn lights on after dark.

Another place where a light sensor may be useful is to sense the position of some object in a situation where it is practically impossible to make any physical contact with the object itself. This would be the case with a continuously revolving object.

One of the handiest light sensing devices available to the experimenter is the photovoltaic solar cell. Because of the recent energy shortage much attention has been paid to the various ways to convert sunlight into electricity. This development, together with efforts in connection with the space program, has led to the production of many low priced solar cells.

Figure 2-14 shows a sketch of a typical selenium solar cell. It is a little more than one inch square and has two flexible wire leads. One side of the cell is photosensitive. The solar cell is rugged in construction and provides a large enough output so that it is not particularly sensitive to noise.

The solar cell produces an output voltage of about 0.5V. This voltage is nearly constant regardless of the amount of light reaching

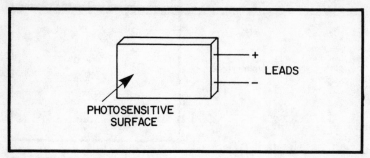

Fig. 2-14. Solar cell.

the cell. Once a little light reaches the cell, the output will rise to about its final value. The output current, on the other hand, depends on the amount of light reaching the cell.

Inasmuch as the solar cell is basically a current generator, it is ideal for use with a transistor which is basically a current amplifier. Figure 2-15 shows the circuit of a light sensitive relay using a solar cell. Note particularly that a germanium transistor must be used in this circuit. The base to emitter voltage of a germanium transistor is only about 0.2V, so the voltage from the solar cell is high enough to make base current flow. A silicon transistor has a base to emitter voltage of about 0.7V and the only way that it will work with a solar cell is if two or more cells are connected in series to get enough voltage to allow base current to flow.

The sensitivity of the circuit of Fig. 2-15 is set by means of the 5K pot, R1. Just about any small signal germanium transistors will work in the circuit. The sensitivity should be set so that the relay will operate at the desired light level.

A variation of the circuit of Fig. 2-15 shown in Fig. 2-16 can be used as a very handy light meter for setting up control systems that respond to changes in light level. This circuit uses a 0 to 1.0 milliammeter as an indicating device. It is possible to calibrate the meter in photometric units, but this calibration is beyond the scope of this book. For control purposes, the light meter can be compared with the photo relay of Fig. 2-15. The indication at which the relay operates can be noted. Once this has been done, the light meter can be used to check any location to see if there is enough light to operate the photoelectric relay.

The photo relay of Fig. 2-15 can also be used to sense the position of an object. Photoelectric sensing is particularly useful with

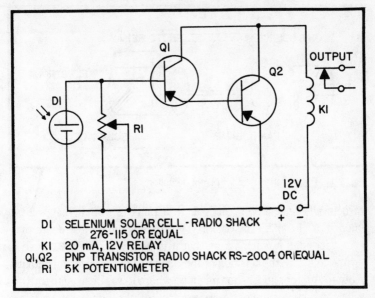

DI   SELENIUM SOLAR CELL - RADIO SHACK
      276-115 OR EQUAL
KI   20 mA, 12V RELAY
QI,Q2 PNP TRANSISTOR RADIO SHACK RS-2004 OR EQUAL
RI   5K POTENTIOMETER

Fig. 2-15. Photo relay using solar cell.

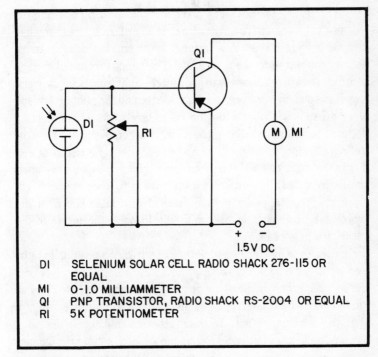

DI   SELENIUM SOLAR CELL RADIO SHACK 276-115 OR
      EQUAL
MI   0-1.0 MILLIAMMETER
QI   PNP TRANSISTOR, RADIO SHACK RS-2004 OR EQUAL
RI   5K POTENTIOMETER

Fig. 2-16. Light meter using solar cell.

41

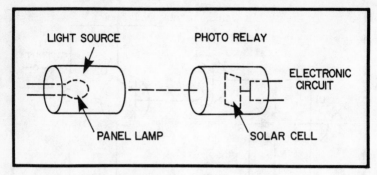

Fig. 2-17. Photo relay used for object detection.

objects where it is very difficult to arrange a switch that will be opened or closed by the object. There are some garage doors that are suspended in such a way that it is nearly impossible to use a switch to sense the position of the door without interfering with its motion. Another application might be where it was desired to turn on a light whenever anyone entered an area through a rather large passageway.

The arrangement of Fig. 2-17 may be used for sensing the position of an object. The light source is an ordinary panel lamp that may be powered either by a battery or by a small step-down transformer connected to the power line. The lamp is mounted inside a small cardboard tube merely to prevent the light from being distracting. If the distance between the light source and the photo relay isn't very great it will usually not be necessary to focus the light. If focusing is necessary, the simplest way to accomplish it is to use the reflector from a flashlight. The photo relay uses the circuit of Fig. 2-15. The solar cell and the electronic circuit are housed in a small cardboard tubes. The reason for the tube is to keep ambient light from reaching the solar cell and providing a false signal. If the ambient light level is high, it will help to paint the inside of the cardboard tube black.

There are instances where the object whose position is to be sensed is not only inaccessible as far as mounting a switch is concerned, but it may be nearly impossible to arrange a photo relay so that the object will interrupt a beam of light when it moves. Figure 2-18 shows a damper that is mounted inside a ventilator. In this case, it would be nearly impossible to arrange a switch so that the damper would contact it, and it would be almost as difficult to arrange a photo relay so that the damper would interrupt a light beam. In the figure the problem is solved by cementing a small mirror to the side of the damper. When the damper is closed, light from the light source will be reflected off the mirror and will hit the photo relay. As soon as the damper starts to open, the mirror will move so that the light will no longer hit the photo relay.

Arrangements of this type are limited only by the ingenuity of the experimenter. With the proper arrangement of mirrors, it is possible to sense the position of almost any object, no matter how inaccessible it might be.

All of the sensors described so far depend on a contact closure or opening to sense the position of the object that is to be controlled. This is by far the most satisfactory type of sensing system for most home control systems. Contacts are reliable and when anything fails in the system it is very easy to troubleshoot. The only real limitation of the use of switches and contacts as sensors is that a separate wire is required for each sensor. In most cases this isn't much of a limitation. Small wires are used for the sensing lines and they are not very expensive. About the only time that this limitation is worth considering is when there are several sensors that are located a substantial distance from the main control point of the system. In such a case it might be worthwhile to multiplex the lines from the sensors. One way of multiplexing these signals is described in the following paragraphs.

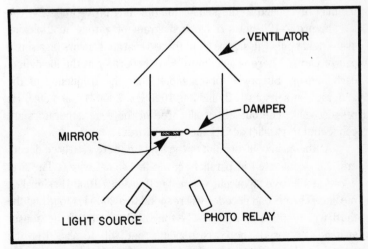

Fig. 2-18. Sensing the position of an inaccessible object.

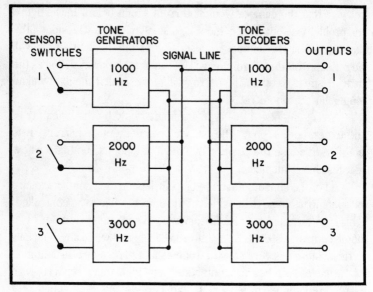

Fig. 2-19. Tone multiplexing system for sensors.

MULTIPLEXING

Later in this book we will describe a method of using audio tones of various frequencies to initiate control action. This same system of tone generators may also be used to multiplex the signals from many sensors onto a single line. We will not describe the details of the tone generators and decoders here, but will merely show how the principle may be applied to the sensor multiplexing problem.

Figure 2-19 shows a block diagram of a tone multiplexing system that puts the signals from three separate sensors on a single pair of wires. There is a separate tone generator at the location of each sensor. Suppose, for example, that the frequency of the generator at sensors 1, 2, and 3 is 1000 Hz, 2000 Hz, and 3000 Hz respectively. The outputs of all three of the tone generators are connected in parallel across the signal wires.

At the main control point of the system there are three decoders, also connected in parallel across the signal wires. The first decoder will have an output whenever there is a 1000 Hz signal on the line. The second decoder will respond to a 2000 Hz tone and the third will respond to a 3000 Hz tone. When any of the sensor switches is closed, the corresponding tone will be placed on the signal line. Thus, if sensor switch 1 is closed there will be a 1000 Hz tone on the line. This tone will be sensed by decoder 1 which will produce an output. The other two sensor switches work in the same way.

The tone multiplexing system has the disadvantage that a separate tone generator and its power supply are required at each sensor location. If the distance between the sensors and the control point is great enough, this may be worthwhile. In most cases it is easier and will cause fewer problems to simply run a separate wire from each sensor back to the control point of the system.

SYNCHRO INDICATORS

Figure 2-20 shows a schematic diagram of a system that can be used to give a remote indication of the position of anything. The proper name for the system is a synchro system, but in the past it has also been called a selsyn (an acronym for self synchronous) system.

The system consists of two devices, called synchros, which look very much like small electric motors. The generator and motor

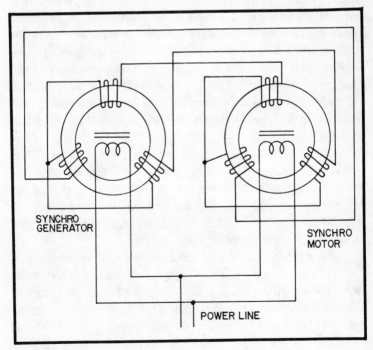

Fig. 2-20. A synchro-system.

are nearly identical in construction. In each case, the stator has three windings that are spaced 120 degrees apart. The rotor has a single winding.

Referring to Fig. 2-20, it is obvious that when an AC voltage is applied to the rotor of the generator, voltages will be induced in the stator windings. The magnitudes and phases of these induced voltages depend on the angular position of the rotor.

In Fig. 2-20 the rotors of the generator and the motor are connected in parallel across an AC source. Note that both rotors are in the same angular position. Under this condition, the voltages induced in the stator of the motor will be the same as those induced in the stator of the generator. As a result, no current will flow in the three wires connecting the two stators. Furthermore, there will be no torque on either rotor.

Now suppose that we turn the rotor of the generator. The voltages induced in the stators of the generator and the motor will no longer be equal. Currents will now flow in the three wires connecting the motor and generator stators. These stator currents will produce magnetic fields. In the motor the field will interact with the field from the rotor in such a way as to make the rotor turn. If there isn't much restraining force on the motor shaft, it will turn in the same direction that we turned the generator shaft. The motor will continue to turn until its rotor takes the same angular position as the rotor of the generator. Thus a synchro system consists of two small devices that resemble electric motors and are connected together by five wires. When the shaft of the generator is turned, the shaft of the motor will follow it.

Figure 2-21 shows an application of a synchro system. The generator is connected through a cord to an overhead garage door. When the door opens or closes, the cord will cause the shaft of the synchro generator to turn. Say, for example, that when the door moves from fully open to fully closed, the shaft of the generator makes 10 complete revolutions.

The synchro generator is connected through five wires (as in Fig. 2-20) to a synchro motor at the control point. The shaft of the motor is connected through a 40 to 1 gearing system to a pointer. When the door moves from fully closed to fully open, the generator will make 10 complete revolutions and the motor will follow it. Because of the 40 to 1 gear ratio, the pointer attached to the synchro

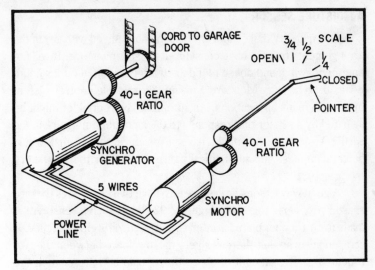

Fig. 2-21. Application of a synchro system.

motor shaft will move through an angle of 45 degrees. A scale can be provided to show just how far open the door is at any time.

The synchro system has the advantage that it can be used to provide a precise, continuous indication of the mechanical position of anything that can be attached to the generator shaft. Because of its construction, a synchro has little wear and will certainly have a much longer life than a potentiometer used in a similar application.

The synchro system also has many disadvantages. In the first place, we need five wires to transmit the position information. Probably more important, we need the full rotor voltage, often 120V, at the control point. This defeats the advantage of using low voltages as we have in most of the other arrangements described in this book. Fortunately few control systems require a sensor with the accuracy and precision of a synchro system.

New synchro motors and generators are probably too expensive for use in a home control system. Many different synchro devices are available in salvage stores that handle military and industrial surplus items. Many of these synchros are rated at 120V, 60Hz and can be used directly.

Some synchros, particularly those used in aircraft applications, are rated for use on 400 Hz lines. A unit of this type will burn out if it is used on a 60 Hz line; however, some such units will function at 60 Hz at reduced voltage.

A MOISTURE SENSOR

One quantity that cannot conveniently be sensed with any of the arrangements that we have considered so far is moisture. It will not operate a switch and doesn't lend itself to operation of either photo relays or synchros. Moisture sensing is a valuable control system function in many applications. An automatic lawn sprinkler might be actuated by a sensor that responds to the amount of moisture in the earth. A storage area might have a moisture sensor that would detect dampness and automatically turn on a heater that would dry the area.

A good way to sense moisture is to have two electrodes that are spaced very close together. Figure 2-22 shows such a sensor. It is made from a copper plated printed circuit. The foil is either etched or scraped away so that there is a very narrow space between the two conductors. The amount of space that should be left between the two plated electrodes depends of the desired sensitivity of the detector and the type of material that the board is made of. It is a good idea to start by etching or cutting a very narrow line in the foil. If the detector is too sensitive, the line can be widened.

When the moisture sensor is dry, it has a very high resistance. As it becomes moist, its resistance decreases but is often in the order of several hundred thousand ohms.

A simple detector circuit that can be used with a moisture detector is shown in Fig. 2-23. Here two NPN transistors are connected in a Darlington arrangement that will provide a very high

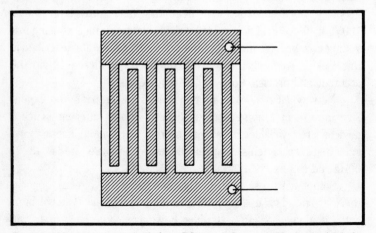

Fig. 2-22. Moisture sensor made from PC board. Copper foil is shown shaded.

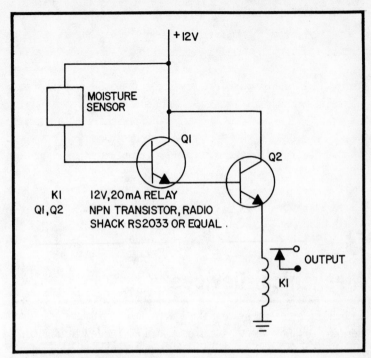

Fig. 2-23. Circuit using moisture sensor of Fig. 2-22.

current gain. Even a very small current through the moisture detector will cause base current to flow in transistor Q1. This in turn will cause even more base current to flow in transistor Q2. The result is that a very small amount of current through the moisture detector will cause the relay in the emitter circuit to operate.

3
Mechanical Devices

In building a control system the electronics technician rarely has serious problems with the electronic portion. The few troubles that develop are usually ironed out with normal troubleshooting procedures. Difficulties usually arise in connection with the mechanical portion of the system. In fact, if it weren't for the mechanical problems involved in controlling things around a home, there would be many more remote control systems in use. Computer enthusiasts. particularly, often plan to use their computers to control various things around the home. Few of these plans are ever realized because of the mechanical, not electrical, difficulties in getting a control system operating.

Another problem in connection with the mechanical portion of a control system is the availability of the mechanical components. If a technician doesn't really know the mechanical requirements of a system it's hard for him to adapt existing mechanical devices to the purpose. Actually most of the mechanical devices needed to control almost anything in a home can be obtained either at a hardware store or from an automobile junkyard. Of course, to be able to select these devices, you must know what is required to do the job at hand. In this chapter we will first review the basic principles of mechanical devices. We will then go on to consider specific examples that can be used for control purposes.

Before we can design a control mechanism, we must know:

1. What is to be moved?
2. How far must it be moved?
3. How fast?
4. How much power is needed to move it?

To answer these questions we must be able to express mechanical quantities in numbers. We may even need to make a few simple calculations. To do this, we must be familiar with a few basic mechanical units.

A unit is used to distinguish one quantity from another. For example, we can't make the statement that a rod is three long. No one would know what we were talking about. If, on the other hand, we were to say that the rod were three feet long, everyone would know exactly what we meant. One of the quantities that we have to deal with in mechanisms is length or distance. We will use the foot as the unit of length. Another quantity that we have to deal with is force. To keep things on familiar ground we will use the pound as the unit of force. With these two units we can now describe another quantity which is work or energy.

Figure 3-1 shows a ten pound weight. We know intuitively that to lift this weight we must exert a force of ten pounds in an upward direction as shown in the figure. The next question is how much energy or work is required to lift the weight. To answer this question we must know how far we're going to lift the weight. More energy is required to lift it over a greater distance than over a smaller distance.

Energy and work, which as far as we're concerned amount to the same thing, are measured as the product of the force exerted and the distance over which some object is moved. The unit of work is

Fig. 3-1. Force is measured in pounds.

51

the foot-pound. Thus if we want to lift our ten pound weight a distance of two feet, the energy required, or the work that is done, is 2 × 10 or 20 foot-pounds.

Energy is a very useful concept in the design of mechanisms. There is a very handy principle in physics that is called the law of conservation of energy. Simply stated, this law says that energy can neither be created nor destroyed; it can merely be converted from one form to another. In terms of a mechanical device this means that whatever energy is put into the device must come out of it. Knowing this we can calculate forces, distances, and so forth in even the most complex mechanisms.

There is another mechanical quantity that we will have to deal with and that is *power*. Power and energy are not the same thing. Power is simply the rate at which work is done or energy is expended. It stands to reason that if we were to lift our ten-pound weight a distance of two feet the power required would depend on how fast we lifted it. Obviously it would take more power to lift the ten-pound weight a distance of two feet in, say, one second then it would to lift it the same distance in one hour.

The simplest way to express power is in terms of foot-pounds per second. Thus if we lifted our ten pound weight a distance of two feet in two seconds, the power would be:

$$\frac{10 \text{ pounds} \times 2 \text{ feet}}{2 \text{ seconds}} = 10 \text{ foot-pounds/second}$$

The foot-pound per second is a very handy unit for computing the amount of power required to move something, but there are two other units of power that are in common use. Electrical power is usually given in watts, and motors are usually rated in horsepower.

The chart in Table 3-1 can be used to convert between foot-pounds per second, watts, and horsepower. The calculations are simple, involving nothing more than multiplications. With a pocket calculator, the task is very simple.

We have now described all of the mechanical units that we will ever be called upon to use in designing or selecting a mechanism for control purposes. The two basic units that we first introduced were force and length. We've found that their product was a measure of work or energy. The final unit that we discussed was power which is the rate at which work is done.

Table 3-1. Conversion between Units of Power.

MULTIPLY THE NUMBER OF → ↓ BY ↘ TO GET	WATTS	HORSEPOWER	FOOT-POUNDS PER SECOND
WATTS	1	746	1.356
HORSE POWER	0.00134	1	0.0018
FOOT-POUNDS PER SECOND	0.7376	550	1

The correct name for a mechanical device that transforms a mechanical force in either magnitude or direction is a *machine*. A machine is a device by which the amount of, or mode of application of, a force is changed for the sake of gaining some practical advantage. A simple way to understand a machine is based on the law of converation of energy. The amount of work *done on a machine* must be equal to the amount of work done *by the machine*. For now we will disregard losses in the machine due to friction. The ratio of the force exerted by a machine to the force applied to it is called its *mechanical advantage*.

LEVERS

A simple lever is a machine. It is usually used to reduce the amount of force required to produce some mechanical motion. The advantage gained by using a lever is called leverage—a term that has found its way into our everyday vocabulary.

Figure 3-2 shows an example of a lever. We are interested in three points on the lever:

1. The point where force is applied, called the *input*.
2. The point where work is done, such as a weight being lifted, called the *output*.

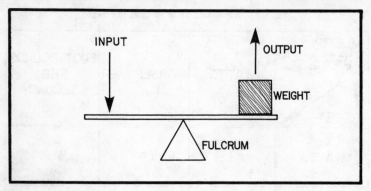

Fig. 3-2. Simple lever.

3. The point, called the *fulcrum*, where the lever is supported.

There are two lengths along the lever that are used to determine the amount of mechanical advantage, or leverage that is gained. These are:

1. The *power arm*, or *force arm*, which is the distance from the fulcrum to the input, that is, the point where force is applied to the lever, and
2. The *load arm*, or *weight arm*, which is the distance from the fulcrum to the output—the point where work is done by the lever.

These two lengths are labeled in Fig. 3-3.

In Fig. 3-4 we have added some numbers to the simple lever of Fig. 3-2. Here at the input point on the lever we are exerting a 2-pound force in a downward direction. At the output, a force of 10 pounds is being exerted in an upward direction. The mechanical advantage in this case is obvious; it only takes 2 pounds of force to lift a 10 pound weight.

The mechanical advantage (or disadvantage) of a lever is easy to calculate. It is simply the ratio of the length of the power arm to the length of the load arm. In this case, the power arm is 5 feet long and the load arm is one foot long. The advantage is then

$$\frac{\text{power arm}}{\text{load arm}} = \frac{5}{1} = 5$$

That is, we will get 5 times as much force out of our lever as we put into it. At first this looks as though we're getting something for

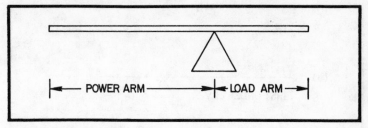

Fig. 3-3. Power arm and load arm of a simple lever.

nothing. We're not. A little reflection on the law of conservation of energy will show that we must put as many foot-pounds of energy into the lever as we get out of it. For this to be true, we must move the input 5 times farther than the output moves. Thus the price we must pay for reducing force is an increase in the distance we have to move the lever.

In Fig. 3-4 we have accomplished three things by using a lever:

1. We reduced the force required to lift a weight.
2. We moved the lever a greater distance than the weight moved, and
3. We reversed the direction of motion.

Although the most obvious reason for using a lever is to reduce the amount of force required to move something, there are other good reasons to use a lever in a mechanism. We may, for example, wish to reduce the distance over which the input to the lever must travel. Sometimes we may use a lever merely to reverse the direction of a force.

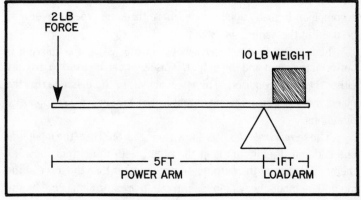

Fig. 3-4. Lever with a mechanical advantage of 5.

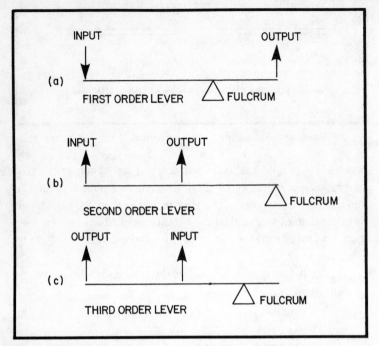

Fig. 3-5. Orders of levers.

ORDERS OF LEVERS

The lever of Fig. 3-3, with the fulcrum between the input and the output, is called a *first order lever*. There are two other orders of levers with different arrangements of input, output, and fulcrum. All three arrangements are shown in Fig. 3-5. The order of lever that we use in a particular application depends on what we want to accomplish. In many applications, more than one type of lever may be used in the same mechanism.

In the first order lever, shown in Fig. 3-5a, the fulcrum is between the input and the output. This lever can be used to magnify either the force required to move an object, or the distance that the object moves. The input and output forces will always be in opposite directions.

The *second order lever* is shown in Fig. 3-5b. Here the input is at one end and the fulcrum is at the other, with the output between them. The force at the output is always greater than the force at the input. The input always moves through a greater distance than the output. The input and output forces are in the same direction.

Figure 3-5c shows a *third order lever*. The output is at one end and the fulcrum is at the other with the input between them. The force at the output is always less than the force at the input, but the output moves through a greater distance than the input. The input and the output forces are always in the same direction. This type of lever is used to magnify the distance of travel at the expense of force.

In all of the examples of levers shown so far we have shown the lever moving in a vertical plane with the resultant force being exerted in an upward direction. This doesn't have to be true. We have used this arrangement merely for convenience. In practice a lever may be mounted in any plane at all.

Most mechanisms operate on the lever principle. In many cases there are many levers linked together, but the principle of operation is the same. Later in this chapter we will see how simple levers can be used to make something move in a particular direction. There are a few other things that we should consider first.

THE PULLEY

Arrangements using pulleys or wheels and cords are very well suited for home control applications. Usually a cord and pulley can be lashed up rather easily, whereas a more complex mechanism would require some special machining. The pulley is a very simple arrangement, but it can do quite a bit. The most obvious application is

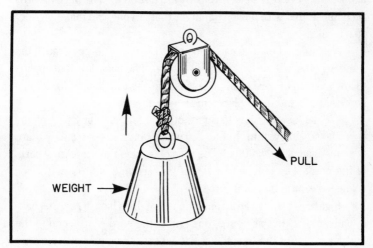

Fig. 3-6. Simple pulley to reverse direction of force.

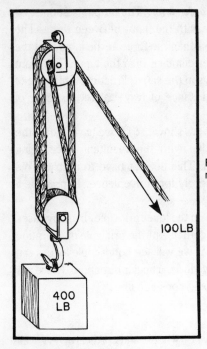

Fig. 3-7. Using pulleys to gain mechanical advantage.

100LB

400 LB

to simply change the direction of a force. In Fig. 3-6 a simple pulley and cord is used to change the direction of a force. Here, we have some type of mechanism that provides a pull in a downward direction and we wish to exert a pull in the upward direction. The pulley lets us accomplish this with an arrangement that anyone can make without special skills.

A system of pulleys can be used to magnify a force in much the same way that a lever will magnify a force. The basic law that applies to any system of cords and pulleys is that the tension, or force, in a cord is the same at all places in the cord.

We can use this law to figure out the forces in any part of any cord and pulley system. In Fig. 3-6 which only uses one pulley there is no mechanical advantage. The tension will be the same in all parts of the cord. The only advantage of this arrangement is that it will change the direction of a force.

Figure 3-7 shows an arrangement that does have a real mechanical advantage. In this figure, one end of the cord is attached to the bottom eyelet of the top pulley. From here, the cord passes over a pulley on the bottom sheave, then passes over a pulley in the top sheave, then back over another pulley in the bottom sheave, and finally over another pulley in the top sheave and out to where the force is applied to it. Invoking our law that states that the tension is the same in all parts of the cord, we can see that the tension will be 100 pounds in all parts of the cord. Inasmuch as the load which is attached to the bottom sheave is supported by four strands of the cord, it will have an upward force of 400 pounds. Thus we can use this arrangement to magnify a force by four. (Of course there will be some loss due to friction on the various pulleys, but there will still be a significant magnification of force.) This arrangement could be used, for example, to lift a 400 pound weight with an applied force of only 100 pounds.

ROTARY MOTION

The most common source of power for a home control system is the electric motor which will be treated in detail in Chapter 5. The electric motor always produces rotary motion and usually it turns much too fast for any practical control application. Of course, some speed reducing mechanism that was originally intended for some other application may be salvaged and used in a control system, but it is often necessary for the builder to fabricate some sort of speed reducing mechanism. A pulley and belt arrangement is often quite useful for this purpose.

Figure 3-8 shows a pulley and belt arrangement that is used to couple power from a motor shaft to another shaft. The pulley at the motor end of the belt has a small diameter, whereas the pulley on the other shaft has a much larger diameter. The result is that the motor shaft must make several revolutions for each revolution of the other

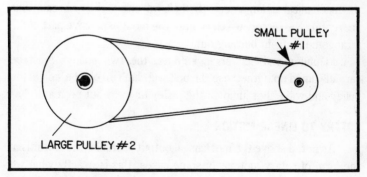

Fig. 3-8. Using a belt and pulleys to reduce speed of rotation.

shaft. Thus the speed is reduced considerably. The amount of speed reduction that can be obtained is inversely proportional to the diameter of the two pulleys. That is

$$\frac{\text{RPM-1}}{\text{RPM-2}} = \frac{\text{D-2}}{\text{D-1}}$$

where
RPM-1 is the speed of pulley 1,
RPM-2 is the speed of pulley 2
D-1 is the diameter of pulley 1, and
D-2 is the diameter of pulley 2.

For example, if pulley 1 were 3 inches in diameter and pulley 2 were 9 inches in diameter, the shaft connected to pulley 2 would only turn at one third the speed of rotation of the motor. Arrangements of this type can sometimes be salvaged from old-fashioned or industrial clothes washers.

When using the pulley and belt arrangement of Fig. 3-8, it is a good idea to have the slack side of the belt on the top. This increases the angle of contact with the drive pulley and thus reduces the tendency of the belt to slip.

This is accomplished by having whichever wheel is supplying the force to turn in such a direction as to pull on the lower side of the belt.

A pulley and belt is a good way to transmit mechanical power in a control system for several reasons. Because of the flexibility of the belt, the arrangement will absorb shocks and smooth out the force applied to the motor. Another advantage is that if something jams in the mechanical arrangement, the belt will either slip or break, thus preventing the motor, which is often the most expensive part of the arrangement, from burning out.

In pulley and belt arrangements, the two pulleys must be carefully lined up. Otherwise the belt will climb up on one side of the pulley and will either jump off the pulley or wear out prematurely.

ROTARY TO LINEAR MOTION

As noted above the mechanical output of an electric motor is in the form of rotary motion. In some cases, this is exactly what we want in a control system. In other cases, what we need is linear, or

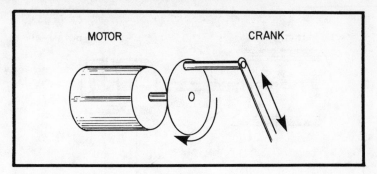

Fig. 3-9. Converting rotary to linear motion.

straight line motion. In such cases we must have a way to convert the rotary motion of the motor into straight line motion.

Probably the simplest device for converting rotary motion into linear motion is the crank shown in Fig. 3-9. Here, as the shaft turns, the rod connected to the crank will move back and forth. Of course, in this case, the speed of the shaft that drives the crank must be low enough to allow the driven member to move smoothly at a reasonable rate of speed. In general, things that are operated by a control system must move slowly. First of all, the power required, as we have seen earlier, increases as things are moved faster. Secondly, anything that moves too rapidly may well be a hazard. For example, a door that swings open very rapidly might be much more of a hazard than a convenience in a home.

Another arrangement that may be used to transform rotary motion into linear motion is the capstan or windlass shown in Fig. 3-10. The capstan is merely a drum about which a cord is wrapped tightly. As the drum turns the cord will move linearly. Arrangements of this type are rather easy to fabricate with few tools. A variation of the same arrangement is the spool shown in Fig. 3-11. Here one end of a cord is secured to a spool that is mechanically coupled to the shaft to a motor or to a shaft where the motor speed has been reduced. As the motor turns, the cord will wind up on the spool. When the motor is reversed, the spool will turn in the opposite direction allowing the cord to unwind.

INERTIA

One thing that is often overlooked in home control systems is the fact that any mechanical object that is to be moved has: resist-

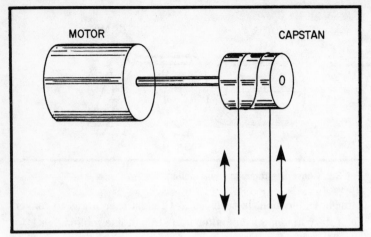

Fig. 3-10. Using a capstan to produce linear motion.

ance to motion, action or change (inertia). The amount of force required to move an object which is at rest, if we ignore the effect of friction, is given by the equation

$$\text{FORCE} = m \times a$$

where m is the mass of the object and a is the acceleration. The mass can be thought of as a measure of the weight of the object. The acceleration is simply how fast it speeds up when we move it. It isn't important to solve the equation; we can simply use it as a guide to give us a little insight into the problems of moving things. The equation says that the amount of force required to move an object increases with the weight of the object and is greater as we try to move the object faster. This is why it is much easier for a man pushing an automobile to keep it moving, than to get it started to move.

As far as control systems are concerned, the above equation says two things. First, if an object is at rest, it takes a lot of force to get it started moving. Secondly, the faster we try to get the object moving, the more force will be required. If this simple rule is ignored, cords will break, motors will stall and burn out, and the system will fail. In most home control systems there is no great hurry to get something to happen. If a door moves from fully closed to fully open in about five seconds there is no harm done. A smaller motor can be used and there is less strain on the whole mechanical system.

PRACTICAL MECHANISMS

Now that we have gone over most of the principles that we will need to construct or adapt a mechanism to a control problem, we need to combine these principles in such a way as to solve a particular problem. To some extent this requires a great deal of ingenuity because many control situations tend to be unique. There are few problems, however, that are typical of any control system.

Probably the best approach to getting something to perform a particular function such as opening and closing a door is to find a mechanism designed for some other purpose that can be adapted to the particular problem. We'll talk more about this in Chapter 15.

The best approach to take in solving any problem is usually the simplest. Figure 3-12 shows a simple arrangement for opening and closing a door. Here two cords are attached to the door in such a way as to pull it open and closed. This approach would work best on an inside door where we have access to both sides of the door. The ends of the two cords are attached to a spool which is driven by a motor. When the motor turns in one direction one cord is wound on the spool and the other is unwound. When the motor turns in the opposite direction, the process is reversed.

Notice that there are springs connected in each of the cords. If the mechanism were perfectly designed, the springs would probably not be necessary, but most practical systems are far from perfectly designed. There is usually some error in the dimensions and placement of the various components of the system. The springs will stretch when necessary to make up for any slop in the dimensions. With springs such as this the arrangement will usually work reasonably smoothly.

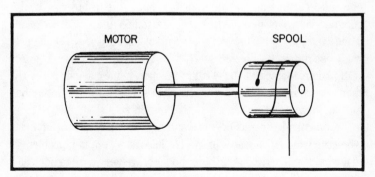

Fig. 3-11. Cord fastened to a spool.

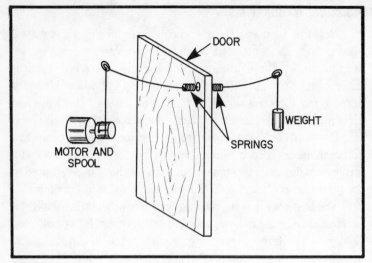

Fig. 3-12. Power mechanism for door.

The arrangement of Fig. 3-12 is not particularly attractive, but it is reasonably easy to get working properly. Sometimes it is easier to use a little ingenuity to disguise the unattractive appearance of a simple mechanism than it is to design an attractive mechanism. Often a clever arrangement of a drapery will completely hide a door opening mechanism.

The chief limitation of the arrangement of Fig. 3-12 is that we need access to both sides of the door. It is only suitable for use on an inside door. It usually isn't advisable to have any part of the door controlling mechanism on the outside where it will be exposed to weather.

The best way to control a door where we only have access to one side, such as an outside door of a house, is to buy one of the commercially available devices that is used to keep a door closed. Then we only need to use the control mechanism to open the door. It will close automatically.

Sliding doors are usually easy to control if one can gain access to them. Of course, the sliding mechanism should work smoothly before any attempt is made to control the door.

Control of the position of windows isn't particularly difficult when the window operates properly. The most common problem is that windows, particularly in older houses, are very hard to open and close. Adding a control mechanism to such a window is only asking for trouble. Window weights and pulleys are hidden from view and the old adage, out of sight, out of mind often applies.

If the position of a window is to be controlled, the window should first be put in good condition. Old cords should be replaced, the pulleys should be well lubricated and work smoothly, and the sliding part should be cleaned and waxed. Such a window will open and close easily and can be controlled with a mechanism quite easily. Another advantage to overhauling the window is that the window casing must be disassembled and this will usually disclose a place where the mechanism can be located out of view.

With a little ingenuity, a mechanism can be devised to move almost anything. Usually, the preferred route is to use a mechanism that was originally intended for some other use. When this is done, the principles of this chapter can be used to get an approximate idea of the force and power required so that a mechanism will be selected that can stand the load.

4

Hydraulic Systems

At first glance, it may seem out of place to include a chapter on hydraulics in a book devoted primarily to electronic control systems. Most electronics technicians would have a tough time trying to design a hydraulic system, and an even tougher time trying to build it and get it operating properly. The reason for including the chapter is that there are many hydraulic systems that were originally designed for other purposes, but can readily be adapted for use in a control system. A typical system of this type is the hydraulic system used to raise or lower the top of a convertible automobile. In order to use such a system intelligently, the technician must have some idea of how it works.

The main advantage of a hydraulic system over mechanical systems is that it will provide a very large force when powered by a comparatively small electric motor. The force is actually transmitted through a fluid—usually some type of oil. Oil has the advantage that in addition to transmitting force, it will also lubricate the system and will not freeze in cold weather.

PRESSURE

In Chapter 3 we discussed most of the physical quantities that we might need in dealing with the mechanical portion of a control system. In using hydraulic systems we need to be familiar with one other physical quantity—*pressure*. Pressure is not the same as force, although it is closely related to it. *Pressure is the amount of force exerted by a fluid per unit of area.* This, if we measure force in pounds, we would measure pressure in pounds per square inch.

Figure 4-1 shows a sketch of a hydraulic system consisting of two chambers that are connected together by a small pipe. The entire system is closed; that is, there is no way for the hydraulic fluid to leak out. We will always have the same volume of fluid in the system. Each of the chambers, or cylinders, is fitted with a close fitting piston. The object of this system is to apply force to one of the pistons and get a larger force from the other piston.

The operation of a hydraulic system is based on a physical law, known as Pascal's principle, which states that if a closed system is filled with a fluid, the pressure is the same in all parts of the system. The principle also requires that the fluid be incompressible. This means simply that we can't reduce the volume of the fluid by compressing it. All hydraulic fluids are incompressible. As applied to the arrangement of Fig. 4-1, this principle means that if we produce a pressure in the fluid by applying a force to one of the pistons, there will be a force exerted at the other piston.

In Fig. 4-1 we have specified that the area of the piston in the smaller piston is 4 sq. in. Thus if we apply a force of 40 lb to this piston we will produce a pressure in the fluid of

$$\text{Pressure} = \frac{\text{Force}}{\text{Area}} = \frac{40 \text{ lb}}{4 \text{ sq. in.}}$$

$$= 10 \text{ lb/sq. in.}$$

According to Pascal's principle, this pressure of 10 lb/sq. in. will be transmitted throughout the fluid. The walls of the containers can't move, so the pressure will move the piston in the larger cylinder. The force exerted on this piston will be equal to the pressure multiplied by the area of the piston.

In Fig. 4-1 we have specified that the area of the larger piston is 600 square inches. Thus the force exerted on the larger piston must be

Force = Pressure × Area = 10 lb/sq in × 600 sq in × 6000 lb

This simple example shows forces can be increased tremendously in a hydraulic system simply by the relative areas in the two pistons. In this example, the force was multiplied by 150. At first this sure looks like we are getting something for nothing, but a little reflection on the principle of conservation of energy described in the preceding chapter will show that we must pay a price for the magnification of force. The energy that we get out of a hydraulic system can't be any greater than the energy that we put into it. The price that we pay in this case is that we must move the smaller piston much farther than the larger piston will move.

Another way of looking at this is to realize that for all practical purposes the hydraulic fluid is not compressible. That is, its volume will not get smaller when we apply pressure to it. Looking at the figure we can see that because of the larger volume of the larger cylinder, it will take the same amount of fluid to move the piston one inch as it would take to move the smaller piston 10 inches. Thus the hydraulic system will indeed magnify force, but it will not increase the energy in the system. We can't get any more energy out of the system than we put into it. In fact, we will actually get a little less energy out because some energy will be lost in the friction between the pistons and the walls of the cylinders and in the friction between the fluid and the walls of the pipe.

MOTOR DRIVEN HYDRAULIC SYSTEM

The simple hand powered system of Fig. 4-1 can indeed be made practical. In fact, automobile hydraulic jacks work on this principle. For any remote control system, however, we need a system that can be controlled electrically.

Figure 4-2 shows a diagram of a hydraulic system where the energy comes from an electric motor. Of course, a motor could be used to apply force directly to a mechanical load, but most motors have a fairly high speed with a comparatively small force. By using the motor to drive a hydraulic system we can convert the rotational force from the motor into a linear motion and we can magnify the force tremendously.

In Fig. 4-2 the electric motor drives a hydraulic pump that produces a pressure in the hydraulic fluid. The first component in the system is a pressure relief valve. This valve is included to bypass the fluid back into the reservoir if the pressure in the system should become dangerously high as it might if some part of the system were not working properly. The next component is an accumulator. Most

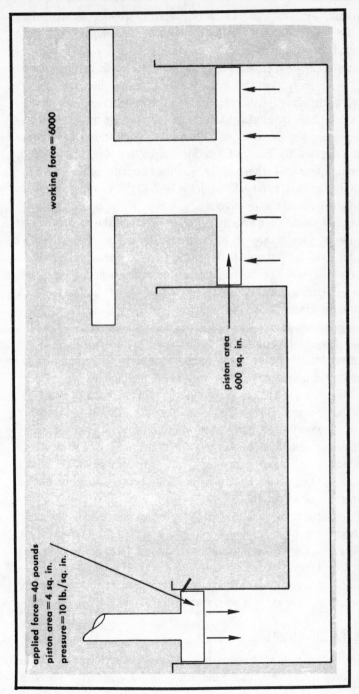

applied force = 40 pounds
piston area = 4 sq. in.
pressure = 10 lb./sq. in.

working force = 6000

piston area
600 sq. in.

Fig. 4-1. Simple hydraulic system.

hydraulic systems include an accumulator to smooth out any pulses in the pressure of the fluid. Sometimes the accumulator will store enough pressure so that the system will start to respond immediately while the motor and pump are still building up pressure.

The delivery line from the pump directs the fluid under pressure to the directional control valve. This valve routes the fluid to one side or the other of a piston in a cylinder depending on which way the piston is to move. With the valve in the position shown in the main diagram the fluid is applied to the bottom side of the piston, forcing it upward in the cylinder. The fluid from the other side of the piston is allowed to pass through the back of the valve to the return line and back to the reservoir.

To force the piston in the opposite direction, the valve is moved to the position shown at the bottom of the figure. Here the fluid under pressure is routed to the top of the piston forcing it downward in the cylinder. The fluid at the bottom of the cylinder now flows through the right hand side of the valve into the return line and back to the reservoir.

Some transfer valves have a neutral position where both of the lines going to the cylinder are blocked by the valve. When the valve is moved to this position, the pump can be turned off and the piston will stay in position being held in place by the fluid on either side.

The simple system shown in Fig. 4-2 is typical of many hydraulic systems. One of its major features is that it is capable of producing very large forces even when powered by a comparatively small electric motor. Many practical hydraulic systems use very high pressures. It is not uncommon to find pressures as great as 1200 pounds per square inch in the hydraulic system used in the steering apparatus of an automobile.

The hydraulic system is great in a home control system where it is necessary to exert a large force for a short period of time. Typical examples include the opening or closing of a very heavy door or window. Often a hydraulic system that was originally used to raise and lower the roof or trunk lid of an automobile can be converted for use in a control system with very little trouble.

HYDRAULIC PUMPS

The purpose of the pump in a hydraulic system is to convert the mechanical energy from the motor into hydraulic energy. We often

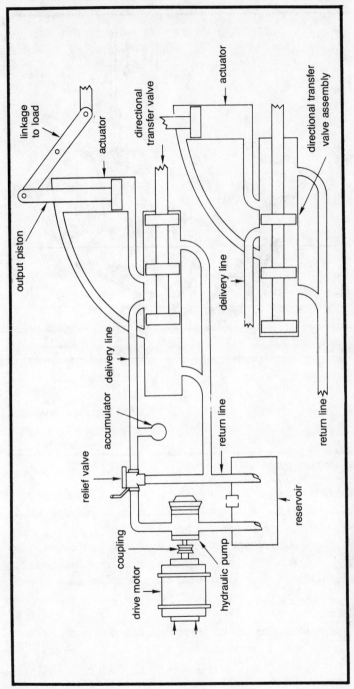

Fig. 4-2. Electrically powered hydraulic system.

71

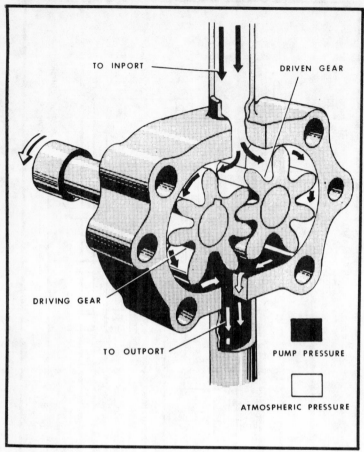

TO INPORT

DRIVEN GEAR

DRIVING GEAR

TO OUTPORT

PUMP PRESSURE

ATMOSPHERIC PRESSURE

Fig. 4-3. Gear pumps.

think of the pump as producing a certain pressure. Strictly speaking this isn't correct. The pump makes the fluid flow and it is restrictions to this flow by the system that produces the pressure.

Many different types of pumping mechanisms are used in the pumps in hydraulic systems and a highly specialized terminology is used to describe them. Basically all hydraulic pumps fall into one of two broad categories—positive displacement and nonpositive displacement.

Considering first the nonpositive displacement pump, it usually delivers a smooth continuous flow of fluid, but it does not have a positive seal to prevent the fluid inside the pump from slipping from the output side back to the input side. As a result if the outlet of a nonpositive pump is completely blocked, the pressure will rise to some maximum value for the particular speed of rotation. When this condition is reached the pump won't do much except churn the fluid around and cause heat.

In the positive displacement pump there is a positive seal against internal slippage. The fluid is usually pumped in pulses as opposed to a smooth continuous flow. If the outlet of a positive displacement pump is completely blocked, the pressure will rise until the driving motor stalls or something breaks. Even with a small motor, this pressure can be extremely high and when converting a hydraulic system for use in a control system, care should be taken not to remove any of the safety devices such as relief valves.

A common type of rotary pump used in hydraulic systems is the gear pump shown in Fig. 4-3. It consists essentially of two gears that are tightly housed in a sealed chamber. The clearances between the teeth of the gears and between the gears and the housing are very small. As the gears turn, some of the fluid becomes trapped in the space between the teeth of the gears and the housing and is forced from the inlet side to the outlet side of the pump. The gears mesh so tightly in the center of the pump that very little fluid can slip between them.

The lobe pump shown in Fig. 4-4 is a variation of the gear pump. It can be thought of as a gear pump where the gears only have three teeth. In the lobe pump there is a better seal in the center where the lobes come together and there is usually more volume between the lobes so that more fluid can be pumped per revolution.

Yet another pump that is used in hydraulic systems is the vane pump shown in Fig. 4-5. The rotor of this pump is slotted and each

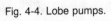
Fig. 4-4. Lobe pumps.

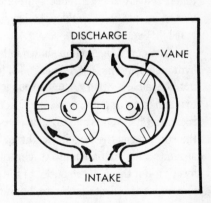

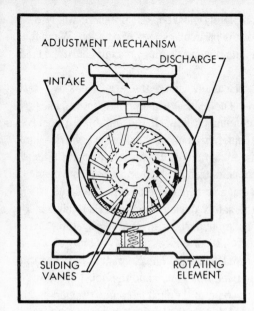

ADJUSTMENT MECHANISM

DISCHARGE

INTAKE

SLIDING VANES

ROTATING ELEMENT

Fig. 4-5. Vane pumps.

slot contains a vane that is free to move in and out. The rotor is not centered in the housing so the vanes will in fact move in and out as the rotor turns. This forms cavities between the vanes and the housing. As the rotor turns the vanes slide in such a way as to make the cavities become larger and then smaller.

The intake of the pump is at the point where the cavities start to become larger. This expansion of the cavities forms a partial vacuum sucking the fluid into the cavities. As the rotor continues to turn, the cavities contract forcing the fluid out through the outlet of the pump.

There are many other types of pumps used in hydraulic systems. Usually any hydraulic system of the type that might be converted for use in a home control system will use one of the pumps that has been described.

The centrifugal pump shown in Fig. 4-6 isn't used very often in hydraulic systems that deliver much force, but we are mentioning it here because it is very useful in applications where water is pumped to water lawn or pump out leaky basements. The pump consists of a rotor that has vanes and turns in an eccentric housing or volute. As the rotor turns the fluid is caught up by the vanes and rapidly spun around. This motion imparts centrifugal force to the water and forces it out through the outlet of the pump. A centrifugal pump will move a lot of fluid at a high rate, but it will not build up much pressure as compared to other types. It is fine, however, for pumping water in applications where there isn't much back pressure.

HYDRAULIC MOTORS

All of the pumps described in the preceding paragraphs are reversible. That is, if instead of applying mechanical energy to the shaft of the pump, fluid is forced through the pump, it will act as a motor and will rotate in the reverse direction. The flow of fluid will be in the opposite direction from that when the pump is acting as a pump. Hydraulic motors are used in such applications in aircraft where it is convenient to deliver hydraulic energy rather than electrical or mechanical energy.

There is little to be gained by using a hydraulic motor in a control system of the type that we are discussing in this book. If rotary motion is required, it is usually much easier to use an electric motor. For this reason we will not say any more about the hydraulic motor.

The hydraulic component that is most useful in a control system is the hydraulic actuator described in the following paragraphs.

HYDRAULIC ACTUATORS

The hydraulic device that provides linear—back and forth—mechanical motion is called an actuator. It consists essentially of a

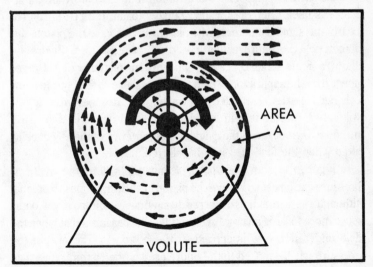

Fig. 4-6. Centrifugal pumps.

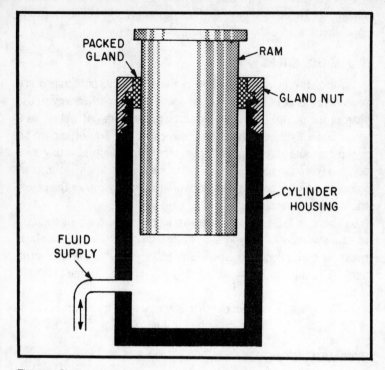

Fig. 4-7. Single acting ram.

cylinder that is equipped with a tight fitting piston. When an actuator is used primarily for lifting heavy loads, the piston rod has nearly the same diameter as the piston. In this case the piston, and sometimes the entire actuator is often called a ram. Figure 4-7 shows a single-acting hydraulic ram. Here the piston or ram is forced upward by the application of fluid under pressure to the bottom of the cylinder. The device is called single acting because the hydraulic fluid will only force the ram in one direction—upward. The ram is returned to the bottom of the cylinder by the weight of the object being lifted when the fluid pressure is relieved.

Figure 4-8 shows a double acting ram where the hydraulic pressure can be used to move the piston in both directions. Note that although the actuator is referred to as double acting, it will apply much more force in an upward direction. The reason is that when the fluid under pressure is applied to the bottom of the piston, the pressure will be applied to the entire area of the piston. When fluid under pressure enters the top of the cylinder, it will only apply pressure to the upper surface of the ram lip which has a much smaller area than the entire ram. Thus although the pressure is the same for upward and downward motion, the force will be much greater in the upward direction.

Although rams may be found in industrial supply houses, the type of actuator that is found much more commonly uses a piston and piston rod. Figure 4-9 shows a single-acting actuator of this type. Here, the piston is made to move to the right in the figure by applying fluid to the left of the piston. When the pressure of the fluid is relieved, the spring at the right of the piston will force the piston to the left side of the cylinder. When the piston is being moved to the right, the force from the fluid must be great enough to not only move the load that is connected to it mechanically, but also compress the spring at the right side of the piston. When the fluid pressure is relieved, the spring will force the piston back to the left side of the cylinder. This type of actuator is used when a high force is required in one direction only.

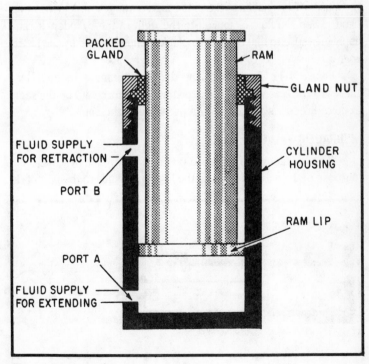

Fig. 4-8. Double acting ram.

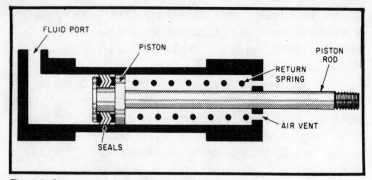

Fig. 4-9. Single acting hydraulic actuator.

Figure 4-10 shows a double-acting actuator. Here the fluid can be applied to either side of the piston so that the full available hydraulic force can be used to move a load in either direction. This arrangement is called an unbalanced actuator. The reason is that it will provide more force driving the piston to the right than to the left in the figure. When fluid is applied to the left side of the cylinder the pressure will be applied to the entire area of the piston. On the other hand, when the fluid is applied to the right of the cylinder, it can't apply pressure to that part of the piston that is blocked by the piston rod.

Figure 4-11 shows a balanced double-acting actuator. Here there is a rod on each side of the piston so the force will be the same in both directions as long as the pressure is the same.

HYDRAULIC VALVES

The operation of a hydraulic system is controlled by valves. The purpose of the valve is to route the hydraulic fluids to the part of the

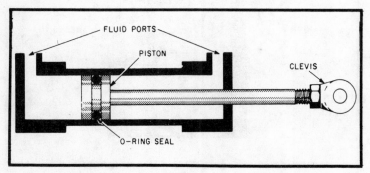

Fig. 4-10. Doubling acting hydraulic actuator.

system where it is required to do the desired work. Hydraulic valves are called by many different names. So far we have called them directional control valves because they are used to fix the direction of flow of the fluid. They are also called by such names as selector valves or power control valves.

Because of the high pressures that are encountered in hydraulic systems, the valves are usually of special construction. Figure 4-12 shows what is called a poppet valve. When the valve is closed, the pressure of the hydraulic fluid will keep the poppet tight against the seat of the valve to prevent it from leaking. When the poppet is moved a very small distance away from the valve seat, the pressure is relieved and the valve can be opened readily.

Probably the most common type of valve used as a directional control valve in a hydraulic system is the spool or piston type of valve shown in Fig. 4-13. It gets its name from the fact that the valving elements look like spools. Note that there are two spools connected to the valve rod. There is a good reason for this. If there were only one spool, the full pressure of the hydraulic fluid would be applied to it and the spool would require a great deal of force to move it. With two spools, the pressure is applied equally to both spools, but the forces are in opposite directions and the net force is zero. As a result the valve can be moved easily even though the fluid might be at a very high pressure.

In some places in hydraulic systems it is desirable to keep the fluid from flowing backward in the system if the pressure is relieved. A check valve is used to prevent this from occurring. Often a check valve is included as a part of a fluid control valve. Figure 4-14 shows a typical spring-loaded check valve. When pressure is applied in the

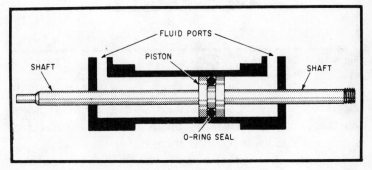

Fig. 4-11. Balanced doubling acting actuator.

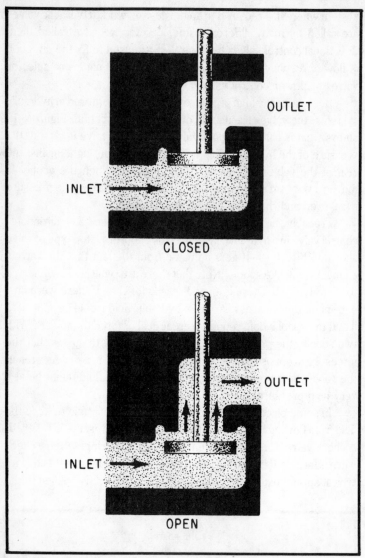

Fig. 4-12. Poppet valve.

direction shown by the arrow in the figure, the ball is forced to the right and the fluid will flow around it . When the pressure is relieved, the spring will force the ball tight against the valve seat preventing the flow of fluid in the opposite direction.

Figure 4-15 shows the type of hydraulic valve best suited for use in a remote or automatic control system. Here a spool type valve

80

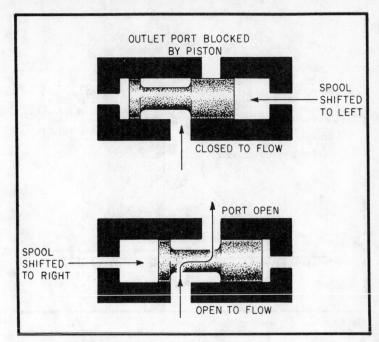

Fig. 4-13. Spool valve.

is opened and closed by a pair of solenoids. When solenoid 1 is energized, the valve body will be pulled to the left in the figure. The pressure will be applied to the right of the piston and the piston will be forced to the left. When solenoid 2 is energized, the process will be reversed and the piston will be forced to the right. When neither solenoid is energized, the valve body will be at the position shown in the figure. Here fluid can neither enter nor leave the actuator so the piston will remain in whatever position it happens to be in.

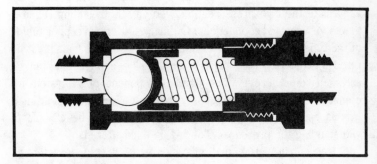

Fig. 4-14. Spring loaded check valve.

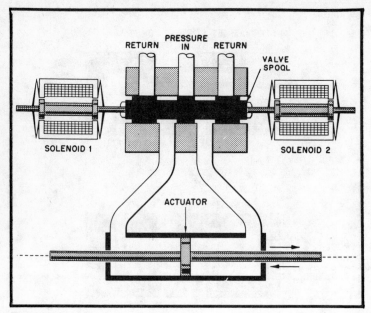

Fig. 4-15. Solenoid operated spool valves.

There are many other versions of the electrically operated hydraulic valve. In some valves a small reversible electric motor is used to move the body of the valve. In any case an electric signal can be used to make the valve move to the desired position. This makes it ideal for use with a control system where the control signals are all electric currents.

HYDRAULIC LINES AND CONNECTIONS

Probably the biggest problem encountered by an electronics technician working with hydraulic systems is due to a failure to appreciate the high pressures involved and the resulting requirements for lines and connections. Ordinary soft drawn copper tubing of the type used in oil burners simply won't do. The lines must be rated for the expected pressures. Usually, enough line can be salvaged from wherever the hydraulic components of the system are obtained. Sometimes it is necessary to extend lines and add fittings. In this case, both the line and the fittings must be capable of withstanding the pressures that will be encountered.

In a control system, hydraulic lines are of two types—semirigid tubing and flexible hose. Most of the lines should be of the semirigid tubing which is rated to withstand the expected pressures. Flexible hose can be used to accommodate motion in the system.

In general, lines should be reasonably straight and short. They should not be exactly straight, however. A straight line will not allow for expansion and contraction and will often result in failure. Figure 4-16 shows the right and wrong way to make lines out of tubing. The bends in the tubing should be neither too sharp nor too gradual.

A bend in a line is measured in terms of the ratio of the radius of the bend to the diameter of the tubing. The correct bend should have a radius of about 2½ to 3 times the radius of the tubing, as shown in Fig. 4-17. Shorter bends will increase the pressure in the line and much larger bends will cause turbulence.

All connections should use flared tubing. The little ferrules that are sometimes used at connections in copper tubing will not withstand the pressure of a hydraulic system. A cross section of a flared

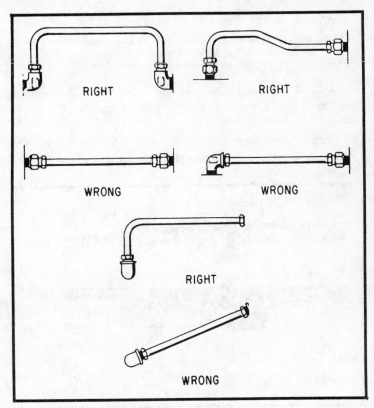

Fig. 4-16. Right and wrong ways of installing tubing.

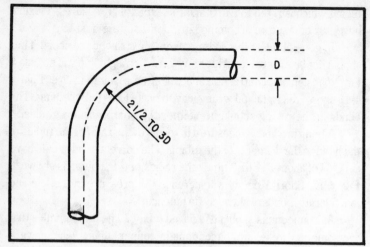

Fig. 4-17. Proper bending radius in tubing.

connection is shown in Fig. 4-18. The connection itself should be of the same material as the tubing. If steel tubing is used, steel connections should be used. If hardened aluminum tubing is used, aluminum fittings should be used.

A flared tube should not be bent or forced into a fitting. Figure 4-19 shows the correct and incorrect ways of making a flared connection.

Similar precautions apply to the use of flexible tubing. The tubing used in a hydraulic system should be capable of handling the

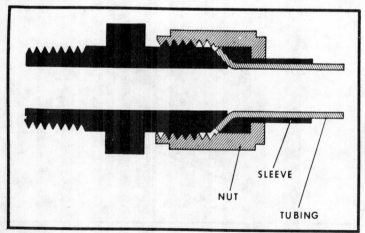

Fig. 4-18. Flared tubing connection.

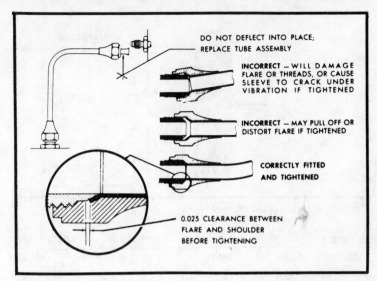

Fig. 4-19. Right and wrong way of making flared tubing connections.

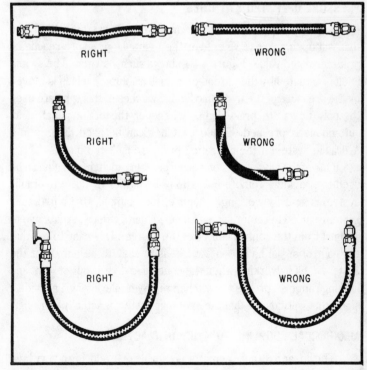

Fig. 4-20. Right and wrong way of installing hose.

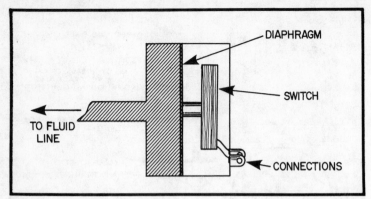

Fig. 4-21. Pressure operated switch.

pressure. The flexible tubing used in automobile power steering systems is of this type. Hydraulic hose tubing can be obtained at industrial supply houses.

Figure 4-20 shows the right and wrong ways of installing hose.

PRESSURE OPERATED SWITCHES

A very useful accessory for any hydraulic system is a switch that will open or close whenever the hydraulic pressure reaches a predetermined value. Figure 4-21 shows such a switch. The switch itself is actuated by the motion of a small diaphragm which is moved by the pressure of the hydraulic fluid. Switches of this type are used to apply power to brake lights whenever the brake pedal of an automobile is pressed. These switches can be used directly in a hydraulic system if the operating pressure isn't too high.

One of the advantages of using a pressure switch is that it can be used to provide a remote indication of whether or not a hydraulic system is actually working. Suppose, for example, that a hydraulic system is used to remotely open a heavy garage door in response to a signal from the control center of the system. It is easy to tell that the control signal has been sent. It isn't easy to be sure that the system itself is working when the controlled door is out of sight and hearing range. A pressure switch in the hydraulic line will provide a signal that will indicate that the hydraulic system is actually working.

MECHANICAL COUPLING TO HYDRAULIC ACTUATORS

The linear hydraulic actuator is capable of producing a very high force along a straight line. In accordance with the law of conservation of energy the price that we pay for the magnification of force is that the piston of the actuator will travel over a comparatively small distance. The actuator may be coupled directly to whatever is to be moved, or the force may be transmitted through levers.

One of the advantages of the hydraulic actuator is that it may often be mounted in a position where other types of application of force would be impractical. Figure 4-22 shows a hydraulic actuator connected to open and close a door. The actuator is placed so that the door will be fully opened when the piston in the actuator reaches its limit of travel. Because the motion of the piston in this particular actuator is limited to about ten inches, the end of the actuator is connected to a point on the door about ten inches out from the hinges. The fixed end of the actuator is mounted on a swivel arrangement so that as the piston moves outward, the door will be forced to open.

The actuator in Fig. 4-22 has two ports for the hydraulic fluid—one on either side of the piston. The actuator is a part of the complete system shown in Fig. 4-23. Here the hydraulic pump is driven by a reversible electric motor which is in turn energized by

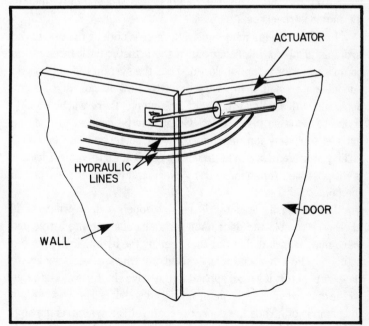

Fig. 4-22. Hydraulic actuator for opening and closing door.

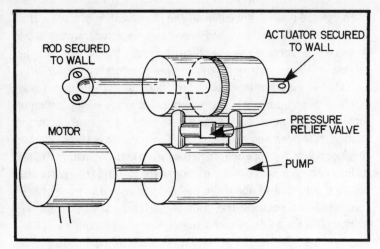

Fig. 4-23. Detail of hydraulic door opener.

the control system. To open the door the motor turns in the forward direction. This pumps hydraulic fluid under pressure behind the piston of the actuator. The fluid in front of the piston is pumped out of the actuator. Thus the piston moves forward without much resistance from the fluid. To close the door, the direction of rotation of the motor is reversed.

In the arrangement of Fig. 4-23, there is nothing to prevent the hydraulic fluid from draining out of the actuator back through the pump if the door is forced closed. Thus the system will not hold an outside door open if the wind tries to blow it closed. In many applications this isn't a problem. For example, there will be nothing trying to close an inside door. Or, it may be desired to open an outside door only long enough to admit someone. The lack of a holding arrangement can be an advantage. Thus if the power fails, it is still possible to open the door by exerting slow steady pressure on the door.

If a hydraulic actuator is used to open a door where it is necessary to hold the door open for an indefinite time, a solenoid valve may be included in the lines feeding the hydraulic fluid to the actuator. The valve is normally closed, but when power is applied to the pump motor, it is also applied to the valve. In this way the valve will be open whenever the pump is operating, but will be closed when the pump is off. Thus the door, or whatever the system is moving, will be held stationary except when power is applied to the system.

CONTROLLING FLUIDS AND LIQUIDS

There are other control operations that involve the control of liquids in addition to hydraulic power systems. For example, an automatic control system might turn on water sprinklers when the lawn is dry, or automatically keep the water in a swimming pool at a constant level. Solar heating systems involve pumping water from one vessel to another and controlling the flow with valves.

The control of fluids such as water is very similar to the control of hydraulic systems. The exception is that water systems will not lubricate the components of the system and water will freeze at low temperatures. For these reasons, pumps and valves that control the flow of water will usually require more maintenance than the corresponding components in hydraulic power system that uses oil as the fluid.

The most common device used to control the flow of water is a simple solenoid actuated valve, like the one shown in Fig. 4-24. In this valve, a spring holds the plunger in the closed position so that when no power is applied the valve will be closed. When power is applied to the solenoid, the plunger will be lifted allowing water to flow through the valve.

When an electrically operated valve is used to control the flow of water, great care must be taken to assure that the valve body is well grounded and that the lines supplying power to the valve are

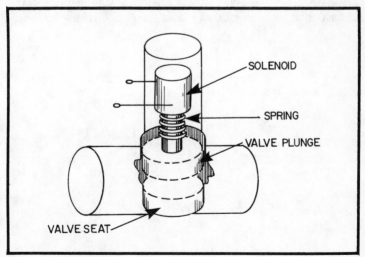

Fig. 4-24. Solenoid valve for regulating water flow.

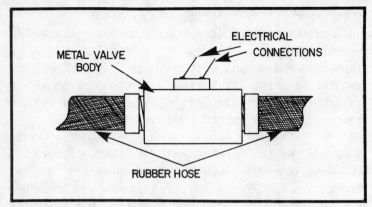

Fig. 4-25. Ungrounded solenoid valve installed in rubber hose.

properly fused. Otherwise an electric shock hazard might exist. This is generally not a problem when the valve is installed in a metal water pipe. Such pipes are invariably well grounded and if there is an electrical breakdown in the windings of the solenoid any leakage currents will be safely grounded. If the lines are tightly fused, the fuse will blow.

A safety problem is more apt to exist when an electrically operated valve is installed between two sections of rubber hose, as shown in Fig. 4-25. Here there is no assurance that the valve body will be grounded and a separate ground wire should be installed.

Sensing the flow of water in a pipe isn't particularly easy. Fortunately it isn't necessary in most simple control applications.

5

Electric Motors and Solenoids

The mechanical and hydraulic arrangements described in the preceding chapters transform mechanical energy. They do not generate it. We still need some source of mechanical motion to produce the mechanical energy. The usual source is an electric motor, or, when only a small amount of motion is required, a solenoid. Selecting a motor for a particular application usually isn't an easy job for the electronics technician; not because it is difficult, but simply because as a general rule, the technician isn't very familiar with motors.

There are many different types and sizes of motors available and selection of the proper motor type for a particular job requires some knowledge of how motors work and what the various specifications mean.

In this chapter we will briefly describe the types of small electric motors that might be used in remote or automatic control systems.

DC MOTORS

Fig. 5-1 shows a functional sketch of an elementary DC motor. It consists of a moving coil called the armature winding and a stationary coil called the field winding. Each of these windings carries a current and therefore produces a magnetic field. Because the armature of the motor rotates, it produces a rotating magnetic field. The field winding being stationary produces a stationary field. It is the interaction of these fields that causes the motor to rotate. In

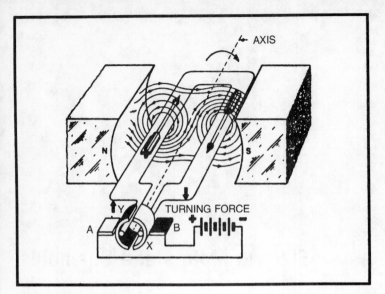

Fig. 5-1. Elementary DC motor.

some motors the field winding may be replaced by a permanent magnetic.

The operation of the motor is shown in Fig. 5-2. At A, in the figure, the armature is in such a position that it produces a magnetic field with a north pole at the top and a south pole at the bottom. Because unlike poles attract and like poles repel, the interaction of the two fields will cause the armature to rotate in a counterclockwise direction.

At B in Fig. 5-2 we have reached a dilemma; the unlike magnetic poles of the armature and the field have lined up. There is no longer any torque to cause the armature to rotate. Unless some other provision was made this is as far as our motor would turn. Fortunately something has been done. The current to the armature is fed through brushes and a commutator. The brush and commutator arrangement reverses the connections to the armature winding at the point in its rotation where the torque would go to zero. This, in turn, reverses the polarity of the field and the armature continues to rotate in a counterclockwise direction.

Our motor of Fig. 5-2 has been simplified to illustrate the basic principle. If a real motor had only one armature winding, the motion would be rather jerky because the torque would be maximum when the unlike poles were coming close together. In a practical motor there are several armature windings, each with its own pair of commutator segments. This produces a practically constant torque. Figure 5-3 shows a sketch of a commutator with several armature windings. As each armature winding is switched into the circuit it will produce a magnetic field that interacts with the stationary field to produce torque.

Our simplified motor has only two magnetic poles from the field, so it is called a two-pole motor. Practical motors may have two, four, or more, poles.

To cause a DC motor to run, a current must flow in both the field and armature windings. The way in which the windings are connected together will have a pronounced influence on how the motor behaves. Before we discuss this aspect of motors, we should get a good understanding of the relationship between voltage, current, torque, and speed in a motor.

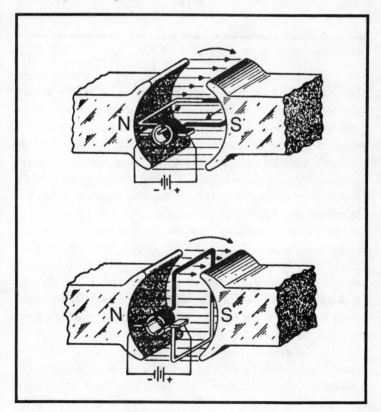

Fig. 5-2. DC motor operation.

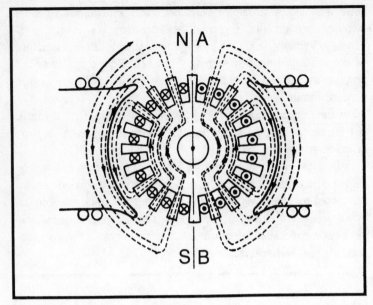

Fig. 5-3. Practical DC motor armature.

A very useful concept in understanding motor operation is that of counter electromotive force or counter emf. This concept is easy to understand when you realize that a motor and a generator are nearly identical. In fact, when a motor is turning, it will generate a voltage in the same way that a generator will. It really doesn't matter whether we turn the motor by rotating its shaft or whether we cause it to rotate by applying a voltage. In either case a voltage will be generated across the armature winding. This voltage opposes the applied voltage. That's why it's called a counter emf.

We can now draw a rough equivalent circuit of the armature portion of our motor. For convenience, we won't bother with the commutator. We know that the windings will have some resistance, so we show a resistor in series in the circuit of Fig. 5-4. We also know that the motor generates a counter emf, so we will also show a generator in series. For the time being that's all we need to explain the basic operation of the motor.

Now, let's suppose that the motor of Fig. 5-4 is at rest. Since it is not rotating, there will be no counter emf. When we close the switch, current will be limited only by the resistance of the armature windings which is quite low. Thus there will be a large surge of current flowing into the motor. This tells us that when we apply power to a motor we can expect a large starting current, until the motor gets going.

Once the motor starts to spin, the counter emf will be developed. This counter emf opposes the applied voltage so that the current flowing through our equivalent circuit will be reduced. The faster the motor turns, the higher the counter emf, and hence the lower the current.

The counter emf also depends upon the strength of the magnetic field from the field winding. The greater the field strength the higher the counter emf for a given speed of rotation. The applied voltage must always be higher than the counter emf because we must have some current flowing through the motor to keep it from turning. If we increase the strength of the field, the motor won't have to turn as fast to generate the same counter emf. This means that increasing the field strength will *decrease* the speed of the motor. Decreasing the field strength will *increase* the speed of rotation of the motor. At first, this sounds backward—like we are getting something for nothing, but there is more to consider. What we really want to get out of the motor isn't speed, but torque or rotational force. The torque increases with both the current through the armature and the strength of the field.

So far we haven't considered the affect of a mechanical load on a motor. Suppose that a motor represented by our equivalent circuit of Fig. 5-4 is running at full speed with no load other than the friction of its bearings. Intuitively, we feel that the motor wouldn't be drawing much current because it isn't delivering much power. This is true, and the reason is that the counter emf is nearly equal to the applied voltage.

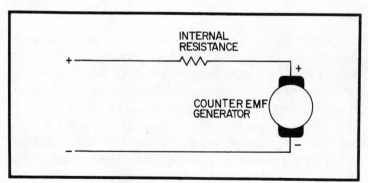

Fig. 5-4. Equivalent circuit of motor armature.

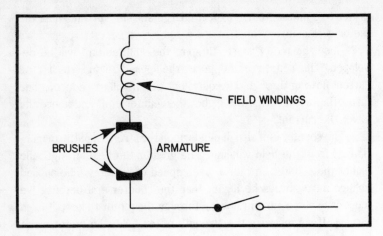

Fig. 5-5. Series motor.

If we now attach to the shaft of the motor some mechanical load, the motor will tend to slow down. This slowing down will reduce the counter emf allowing the current to increase. The increase in current will cause the torque to increase so that the motor can drive the load. Of course, if we apply too much of a load to the motor we may stall it. In the stalled condition there will be no counter emf with the result that the current will be excessive and will probably burn out the armature winding.

There are three basic types of DC motors; they differ primarily in the way that the armature and field windings are connected together. When we select a type of motor for a particular application, we are usually interested in how its speed and torque vary with different load conditions.

THE SERIES MOTOR

Figure 5-5 shows a diagram of a motor with the armature and field windings connected in series. Not surprisingly, it is called a series motor. In this arrangement all of the armature current also flows through the field winding. For this reason the field winding consists of comparatively few turns of heavy wire.

From an earlier discussion of motors, we can see how the series motor will behave with varying load conditions. First let's look at what happens with no load. When the motor is at rest, there will be no counter emf. Thus when the switch in Fig. 5-5 is closed there will be a large current through the armature and field windings which are connected in series. Since the torque increases with both armature current and field strength, there will be a very high starting torque. As the motor turns faster and faster, a high counter emf will be developed, but at the same time it will weaken the field, tending to make the motor turn even faster. With no load a series motor will develop damaging high speeds. They should never be run unloaded.

A mechanical load applied to the shaft of a series motor will slow it down. This will reduce the counter emf so that current and hence the torque will increase. A series motor therefore has a very high starting torque, and a speed that is determined primarily by the mechanical load. It is ideal for applications where a high torque is required for a short period of time. This includes many control applications.

Figure 5-6 shows the characteristics of a series motor in graph form. Of course, this particular motor is much larger than any motor that would find application in a control system, but the shape of the curves will be the same for just about any series DC motor. The curves show that as the load on the door increases, the speed will decrease, and the torque, the current, and the efficiency will all increase.

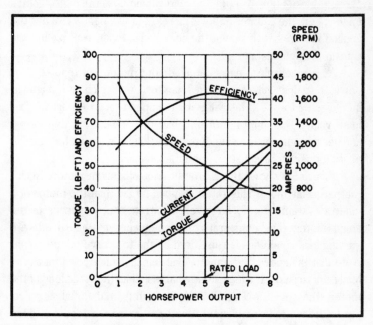

Fig. 5-6. Characteristic curves of a series motor.

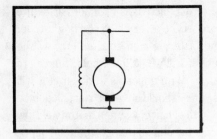

Fig. 5-7. Shunt motor.

THE SHUNT MOTOR

Figure 5-7 shows a motor with the field and armature windings connected in parallel across the power source. Because of this parallel connection it is called a *shunt* motor. The field winding usually has many turns of fine wire. Compared to the series motor, the field current is small but the field strength can be made as large as necessary by using many turns of wire. With this arrangement, the field current, and hence the field strength, is nearly constant.

When power is applied to the motor of Fig. 5-7 the armature current will be high, causing a high torque. This starting torque will not be as high as in the series motor because the surge of starting current doesn't flow through the field winding. As the motor comes up to speed, the counter emf will increase until it nearly reaches the applied voltage. At this point the speed will stop increasing and will remain nearly constant.

As a mechanical load is applied, the motor will tend to slow down, but this will reduce the counter emf which will, in turn, increase the torque. This will bring the speed back to nearly its "no load" value. The shunt motor thus has a reasonable starting torque and nearly constant speed under varying load conditons. It is used in applications where constant speed is required.

Figure 5-8 shows the characteristics of a shunt motor. Again, the particular motor illustrated is much larger than the motor of a control system. It will, however, illustrate the properties of the shunt motor. The comparative independence of the speed with varying load conditions is illustrated by the top curve on the graph. Note that the speed of the motor will remain nearly constant as the load is increased. At a load somewhat less than the rated load of the motor, the speed will start to drop slightly, but for all practical purposes, we can consider the shunt motor to be a constant speed device. Both the current and the torque increase linearly with load, and the efficiency is maximum at some load less than the rated load of the motor.

THE COMPOUND MOTOR

It is possible to use two field windings on a motor—one in series with the armature, and one in parallel with it as shown in Fig. 5-9. This arrangement is called a *compound* motor. There are two possible ways to connect the two field windings of a compound motor. If the two field windings are connected so that they will *aid* each other the motor is called a *cumulative compound motor*. If they oppose each other it is called a *differential compound motor*.

The cumulative compound motor behaves a lot like a series motor. There will be a high starting torque because one of the field windings is in series with the armature. As in a series motor, the speed will increase as the mechanical load decreases. It will not increase without limit, however, because the parallel field has a nearly constant field strength. This motor can be used in applications

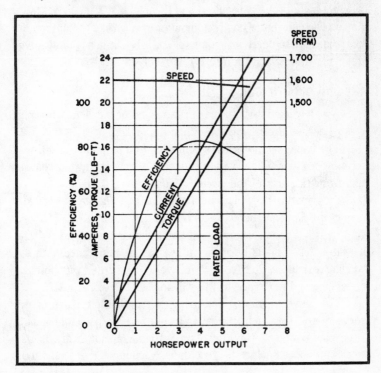

Fig. 5-8. Characteristic curves of a shunt motor.

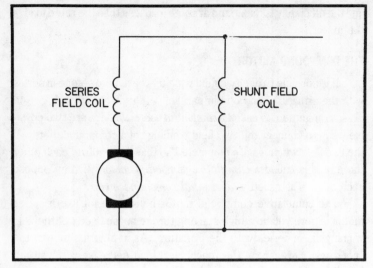

Fig. 5-9. Compound motor.

where a high torque is required and the load may vary over a wide range. The differential compound motor is much like a shunt motor, but it has a more nearly constant speed. It has little advantage over the shunt motor, and therefore it is rarely used.

THE UNIVERSAL MOTOR

A widely used variation of the series motor is called a *universal motor* because it can work on either AC or DC.

The universal motor uses laminated iron parts just as a transformer does. This eliminates the adverse effects of eddy currents in the magnetic parts of the circuit. Usually fewer turns are used in both the armature and field windings so that the reactance will be low enough to allow an adequate current to flow.

Usually a universal motor will run somewhat faster on DC than it will on AC. This is because on DC we are concerned with the winding resistance whereas with AC we are concerned with both the resistance and reactance of the windings.

The universal motor has many of the characteristics of the series motor. It has a high starting torque and is capable of handling large loads over a short period of time. Universal motors are used in small appliances such as electric drills, vacuum cleaners, and food blenders.

REVERSING MOTOR DIRECTION

In most control systems it is necessary to have control over which direction a motor runs. For example, if a motor is used to open a door, it will probably also be used to close the door. It will be necessary for the motor to run in one direction to open the door and the reverse direction to close the door.

The direction of rotation of all the motors described in this chapter can be reversed by reversing the direction of current flow in either the armature or field windings but not both. Figure 5-10 shows how a DPDT relay may be used for this purpose. The relay has a low voltage coil (6-24 volts) so that it can be used with any of the control arrangements described earlier. When the relay is not energized, the motor will run in one direction and when the relay is energized, the motor will run in the opposite direction.

LIMITING ROTATION

There are many applications where we want to limit the amount of travel of a motor controlled device. Usually this isn't important in cases where we can either see what is being moved or have an indication of its state. It can be important, however, when we don't have a good idea of what is going on physically.

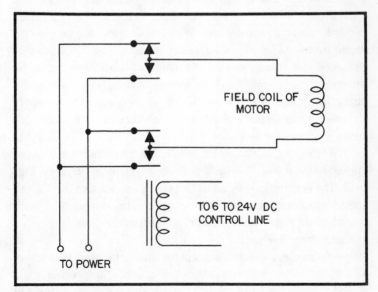

Fig. 5-10. Motor reversing relay.

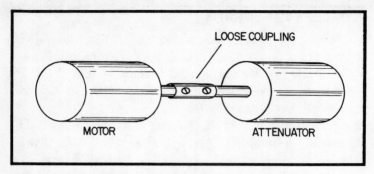

Fig. 5-11. Limiting rotation with a slipping coupling.

For example, suppose we are using a small motor to turn an attenuator that controls the volume of a stereo system. We push a button that makes the motor run forward to increase the volume, and push another button that makes the motor run in reverse to decrease the volume. The human ear is not a good judge of audio volume, so it is possible that we might try to continue to increase the volume after the attenuator has reached the limit of its rotation. The motor used in the system might well be strong enough to completely ruin the attenuator.

One approach to the problem is to make the mechanical coupling between the motor and its load rather loose so that the shaft will slip if the motor drives the load to its limit. In Fig. 5-11 a motor shaft is coupled to an attenuator through a coupling which grips the two shafts by means of two set screws. If one of the set screws is tightened just enough so that the motor will turn the shaft of the attenuator, the coupling will slip when the attenuator reaches the limit of its rotation. The principal disadvantage of this arrangement is that after it has slipped a few times, it may become erratic and not turn the attenuator at all.

Another approach to the problem is to reduce the voltage applied to the motor by means of a series resistor as shown in Fig. 5-12. The resistor is adjusted so that the motor will turn the shaft of the attenuator but will stall when the limit of rotation is reached. With a small motor and a series resistor, no harm will be done if the motor is stalled for a few seconds.

A more elaborate arrangement for limiting the amount of travel of a motor driven device is shown in Fig. 5-13. Here relay K1 is a DPDT relay that reverses the direction of rotation of the motor, and

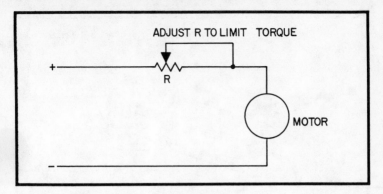

Fig. 5-12. Limiting rotation by limiting torque.

switches S1 and S2 are normally closed, snap-action limit switches. If this circuit were used with our motor driven attenuator, one switch, S1, would open when the attenuator was at the highest volume setting and the other switch, S2, would open at the lowest volume setting.

When power is applied to the circuit of Fig. 5-13, the motor will start to run in the forward direction. As soon as the attenuator reaches the high-volume limit of its rotation, switch S1 will open.

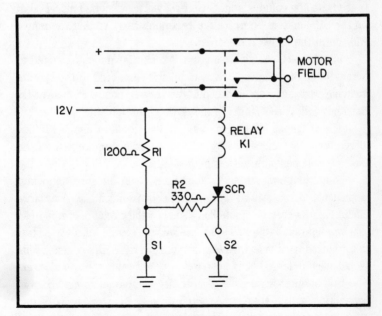

Fig. 5-13. Electronic motor reversing circuit.

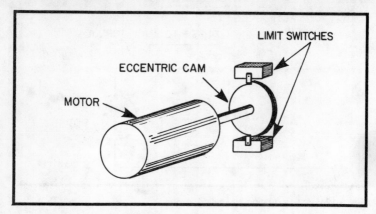

Fig. 5-14. Cam and limit switch to limit motor rotation.

This will allow gate current to flow into the SCR. The SCR current will energize the reversing relay K1 so that the motor will immediately reverse direction before the attenuator is damaged.

Of course, switch S1 will close again as soon as the motor reverses direction. This won't affect anything, however, because the SCR will continue to conduct once it has fired.

Now suppose the motor is allowed to run until the attenuator reaches its low volume limit. Switch S2 will now open. This will shut off the SCR and de-energize the reversing relay so that the motor will run in the forward direction.

Figure 5-14 shows a cam and two limit switches arranged to detect limits of rotation. The system isn't restricted to this type of switching arrangement. The limit switches can be used in any way that they will detect limits of any type of motion.

There are some disadvantages to the circuit of Fig. 5-13. The chief one is that the only way we can reverse the direction of the motor to reduce volume is to run the system up to volume where the reversing relay will be energized. This could be very annoying, particularly if the stereo were capable of producing a very large amount of power. One solution is to modify the circuit of Fig. 5-13 so that when power is applied to the circuit, the motor will always start to turn in a direction as to lower, rather than raise, the volume. This would prevent large blasts of sound while adjusting the sound level.

Still another arrangement is to use the circuit of Fig. 5-13 in connection with a reversing switch or reversing relay. With this arrangement the direction of rotation is controlled by the position of the switch or relay. The limit switches are used only to change the direction if the motor should reach its limit of travel.

Most of the small motors used in control systems will probably be DC motors because it is so easy to reverse them and to control their speed. The principal limitation of the DC motor as far as home control systems is concerned is that it requires DC for operation and the power available in a home is AC. This means that the AC must be rectified for operating the motor. In a small motor this isn't a problem. Rectifier diodes of the types used in electronic power supplies can be used satisfactorily.

In applications where a lot of power is required, it is probably best to use an AC motor. The basic principles of AC motors are covered in the following paragraphs.

AC MOTORS

There are a great variety of AC motors that are available for building control systems. If a new motor is being purchased for a particular application, the manufacture or his agent can help in selecting a motor that is well suited for the application. This makes selecting the motor very simple. Unfortunately, builders of small control systems usually do not buy new motors. They use second hand motors that are purchased from a second hand outlet, or are salvaged from an appliance that has been discarded. In order to make proper use of such motors we need to know something about the various types of AC motors, their characteristics, and how to recognize the different types.

In general, small AC motors can be grouped into the six different categories listed in Table 5-1. The biggest difference between all of these different kinds of motors is in the amount of starting torque that they will develop and in their starting current requirements. Table 5-2 lists many different types of AC motors and their pertinent characteristics.

The reason that the starting requirements of a single phase motor are so important is that a single-phase motor with an armature and field coil will not start by itself. Once it is started, a rotating magnetic field is produced and the motor keeps rotating. Most of the differences between the various types of motors involve the arrangement that is used to get the motor started in the desired direction.

Table 5-1. Types of Small Electric motors.

1. Split-Phase (SP)
2. Capacitor
   a. Capacitor Start (CS-IR)
      (Capacitor Start-Induction Run)
   b. Two-Value Capacitor (CS-CR)
      (Capacitor Start-Capacitor Run)
   c. Permanent-Split Capacitor (PSC)

3. Wound-Rotor
   a. Repulsion-Start (RS)
   b. Repulsion-Induction (RI)
   c. Repulsion (R)
4. Shaded-Pole
5. Universal or Series (UNIV)
6. Synchronous

Figure 5-15 shows a diagram of a split-phase AC motor. It has two field windings. Once it has started to rotate, the main winding will generate a rotating field and will keep the motor turning. The auxiliary winding is used to generate the starting torque. When the motor is at rest, the centrifugal switch is closed switching the auxiliary winding into the circuit. Once the motor develops sufficient speed, centrifugal force causes the switch to open and the motor runs on the current flowing in the main winding. The direction of rotation can be changed by reversing the connections to the auxiliary winding.

The split-phase motor has a low starting torque as compared to other types. It is only suitable for driving loads that have a low starting torque such as fans. Because of the low starting torque and high starting current, motors of this type are usually used where low cost is important.

There are several types of motors that use a capacitor in connection with a winding to obtain starting torque. Figure 5-16 shows a capacitor-start, induction-run motor. This motor is quite similar to the split-phase motor with one very important difference. A capacitor is used in series with the auxiliary winding. The presence of this capacitor in the starting circuit will give the motor about twice as much starting torque as the split-phase type with only two thirds of the starting current. The direction of rotation in this motor can also be reversed by reversing the connections to the auxiliary winding. The principal limitation of the capacitor-start motor is that the capacitance of the electrolytic starting capacitor may be reduced at low temperatures. This, in turn, will reduce the starting torque.

Figure 5-17 shows another capacitor type of motor where one capacitor is used for starting and another smaller capacitor is left in the circuit all the time. The running capacitor provides power factor correction and reduces the required operating current. It has a little more starting torque than the other types and can handle loads that are harder to start. The current requirement is about the same as for the motor of Fig. 5-16. As with the other types, the direction of rotation can be changed by reversing the connections to the auxiliary winding.

Figure 5-18 shows a different type of capacitor motor that doesn't require a centrifugal switch. It is a permanent-split capacitor motor. The capacitor is left in the circuit at all times. Thus it is like the motor shown in 5-17 except that the same value of capacitance is used for both starting and running. The price that we pay for this simplification is that the starting torque is much lower—about the same as that of the split-phase motor.

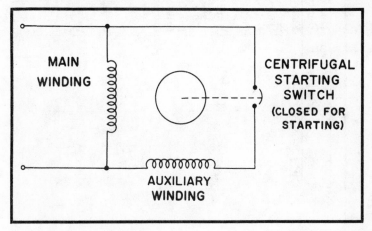

Fig. 5-15. Split phase AC motor.

Table 5-2. Characteristics of AC Motors.

Type	Horsepower ranges	Load-starting ability	Starting current	Characteristics	Electrically reversible
Split-phase	1/20 to 1/2	Easy starting loads. Develops 150 percent of full-load torque.	High; five to seven times full-load current.	Inexpensive, simple construction. Small for a given motor power. Nearly constant speed with a varying load.	Yes.
Capacitor-start	1/8 to 10	Hard starting loads. Develops 350 to 400 percent of full-load torque.	Medium, three to six times full-load current.	Simple construction, long service. Good general-purpose motor suitable for most jobs. Nearly constant speed with a varying load.	Yes.
Two-value capacitor	2 to 20	Hard starting loads. Develops 350 to 450 percent of full-load torque.	Medium, three to five times full-load current.	Simple construction, long service, with minimum maintenance. Requires more space to accommodate larger capacitor. Low line current. Nearly constant speed with a varying load.	Yes.

Type	Horsepower	Starting characteristics	Starting current	Remarks	Reversible
Permanent-split capacitor	1/20 to 1	Easy starting loads. Develops 150 percent of full-load torque.	Low, two to four times full-load current.	Inexpensive, simple construction. Has no start winding switch. Speed can be reduced by lowering the voltage for fans and similar units.	Yes.
Shaded pole	1/250 to ½	Easy starting loads.	Medium.	Inexpensive, moderate efficiency, for light duty.	No.
Wound-rotor (Repulsion)	1/6 to 10	Very hard starting loads. Develops 350 to 400 percent of full-load torque.	Low, two to four times full-load current.	Larger than equivalent size split-phase or capacitor motor. Running current varies only slightly with load.	No. Reversed by brush ring re-adjustment
Universal or series	1/150 to 2	Hard starting loads. Develops 350 to 450 percent of full-load torque.	High.	High speed, small size for a given horsepower. Usually directly connected to load. Speed changes with load variations.	Yes, some types.
Synchronous	Very small, fractional	N/A[1]	N/A	Constant speed.	N/A

[1]N/A = not applicable.

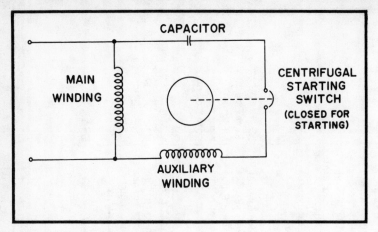

Fig. 5-16. Capacitor start AC motor.

Because no starting mechanism is used in the motor of Fig. 5-18, it can be run at variable speeds by reducing the supply voltage. The speed cannot be reduced below about 75% of the synchronous speed because the torque will drop rapidly and the motor may stall.

All of the motors we have discussed so far have no electric connections from the outside to the rotor. The rotor winding is usually a few turns of heavy wire, often in what is called a squirrel cage configuration.

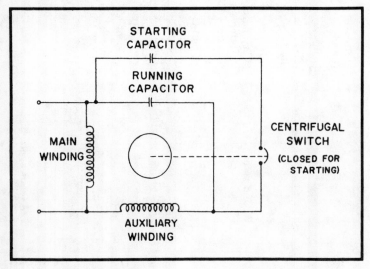

Fig. 5-17. Two capacitor motor.

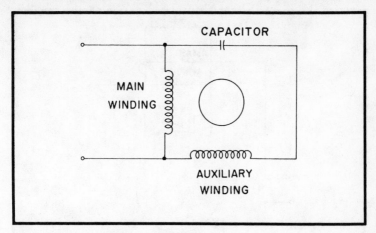

Fig. 5-18. Permanent split capacitor motor.

There are several types of motors that use wound rotors. These motors are more expensive than the split-phase or capacitor types of motors and require more maintenance because the brushes and commutator will wear. Wound rotor motors can provide a very high starting torque at a comparatively low starting current.

The brushes in the repulsion motor are arranged in such a way that the magnetic field of the rotor is inclined with respect to the field from the stator. The result is that there is a strong torque even before the motor starts to rotate. Figure 5-19 shows a repulsion-

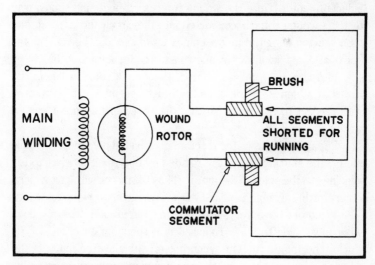

Fig. 5-19. Repulsion-induction motor.

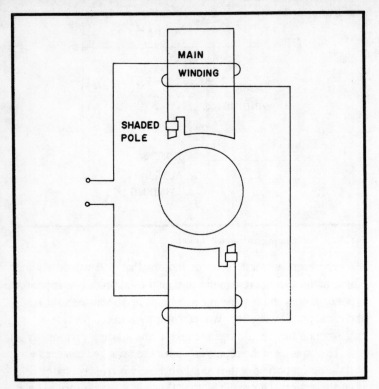

Fig. 5-20. Shaded pole motor.

induction motor. Here the wound rotor is connected in such a way that the wound rotor is connected into the circuit for starting, and then shorted after the motor comes up to speed. Thus the motor starts as a repulsion motor and runs as an induction motor.

In a straight repulsion motor, the wound rotor is left in the circuit at all times. Sometimes the brush assembly can be rotated with respect to the stator so that the speed of the motor can be varied.

Figure 5-20 shows the diagram of a shaded pole motor. In this motor, the starting torque is provided by two small short-circuited windings on the poles of the motor. Thus the motor will start with no other starting arrangement and will always run in the same direction. These motors have very little starting torque and are only used where a very low torque is required as in a phonograph or an electric clock. The motor runs in synchronism with the frequency of the power line so it can be used to drive timing mechanisms accurately.

MOTOR RATINGS

One of the problems in using a second hand motor is the question of whether or not it will do the job. To some extent, particularly with a small motor, the best way to be sure of this is to try it in the application while carefully watching for signs of overloading or overheating.

The nameplate of the motor, which usually looks something like that shown in Fig. 5-21, gives a great deal of information about the motor. The top line on this nameplate gives the specifications of the motor that have been standardized by the National Electrical Manufacturer's Association (NEMA), as well as the manufacturer's identification data. Most of the information on the top line has to do with the shaft size, amount of insulation, and similar factors. Probably the only item of interest is the insulation rating which has to do with the maximum temperature at which the motor can be operated. There are four classes applicable to fractional horsepower motors. They are tabulated below:

Class	Max Temp
A	221 F
B	266 F
F	311 F
H	356 F

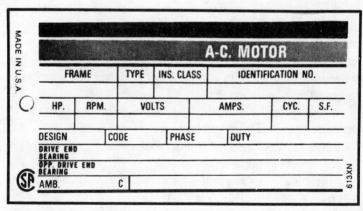

Fig. 5-21. Typical motor nameplate.

113

Naturally a motor can be operated at below its rated temperature, but it shouldn't be allowed to operate where temperature will exceed the rated value.

Most of the entries on the second line of the nameplate are self-explanatory except the Service Factor SF. This is the amount of overload that the motor can tolerate on a continuous basis. The Design entry of the third line applies only to larger motors. The code letter specifies the ratio of the locked rotor to the horsepower rating. In small motors, the code is usually K or L. K means that the locked rotor kva will be between 8 and 9 times the horsepower rating of the motor; L means that it will be between 9 and 10 times.

MOTOR SERVICING AND REPAIR

A new electric motor is a very reliable device and will normally require very little servicing, except for occasional lubrication. A second-hand motor, which is what will probably be used in the majority of control systems, may require servicing to bring it back to good operating condition.

The first thing to do with a salvaged motor is to clean it thoroughly. Dirt and dust should be removed from all of the air passages. Otherwise, the motor can overheat under normal operating conditions.

The bearings must be checked for wear. This is done by moving the rotor shaft sideways and back and forth. If the motor is used in intermittent service, a small amount of side play and end play can be tolerated, but if the movement is excessive, the motor will probably cause trouble—usually low torque, and excessive starting and running current. The motor should be lubricated, preferably in accordance with the manufacturer's instructions if they can be obtained. If not, a small amount of machine oil in the bearings will be better than nothing. Be careful not to use too much oil. It will just lead to the accumulation of dust and dirt.

If the motor uses a centrifugal starting switch, the contacts should be cleaned. This should be done with very fine sandpaper. Do not use emery cloth because emery powder is an electrical conductor and can cause problems.

Table 5-3 lists some of the common problems encountered in AC motors together with their solutions.

Table 5-3. Common Motor Troubles and Repairs.

Cause	Remedy
Motor Fails To Start	
Fuses blown, switch open, broken or poor connections, or no voltage on line.	Check for proper voltage at motor terminals. Examine fuses, switches, and connections between motor terminals and points of service. Look for broken wires, bad connections, corroded fuse holders. Repair or replace as necessary.
Defective motor windings.	Locate and repair.[1]
Motor Hums But Will Not Start	
Starting winding switch does not close.	Clean or replace and lubricate if needed.
Defective starting capacitor.	Replace.[1]
Open rotor or stator coil.	Locate and repair.[1]
Motor overloaded.	Lighten load. Check for low voltage.
Overloaded line or low voltage.	Reduce electrical load. Check wiring. Increase wire size. Notify power company.
Bearings worn so that rotor rubs on starter.	Replace bearings. Center rotor in stator bore.[1]
Bearings too tight or lack of proper lubrication.	Clean and lubricate bearings. Check end bells for alignment.
Burned or broken connections.	Locate and repair.
Motor Will Not Start With Rotor In Certain Position	
Burned or broken connections; open rotor or stator coil.	Inspect, test, and repair.[1]
Motor Runs But Then Stops	
Motor overloaded.	Lighten motor load. Check for low voltage.
Defective overload protection.	Locate and replace.[1]
Slow Acceleration	
Overloaded motor.	Lighten motor load.
Poor connections.	Test and repair.
Low voltage or overloaded line.	Lighten line load. Increase size of line wire.[1]
Defective capacitor.	Replace.[1]

See footnote at end of table.

Table 5-3 cont.

Excessive Heating	
Cause	Remedy
Overloaded motor.	Reduce motor load.
Poor or damaged insulation; broken connections; or grounds or short circuits.	Locate and repair.[1]
Wrong connections.	Check wiring diagram of motor.
Worn bearings or rotor rubs on stator.	Renew or repair bearings. Check end bell alignment.[1]
Bearings too tight or lack of proper lubrication.	Clean and lubricate bearings. Check end bell alignment.
Belt too tight.	Slacken belt.
Motor dirty or improperly ventilated.	Clean motor air passages.
Defective capacitor.	Replace.
Excessive Vibration	
Unbalanced rotor or load.	Rebalance rotor or load.
Worn bearings.	Replace.[1]
Motor misaligned with load.	Align motor shaft with load shaft.
Loose mounting bolts.	Tighten.
Unbalanced pulley.	Have pulley balanced or replaced.
Uneven weight of belt.	Get new belt.
Low Speed	
Overloaded.	Reduce load.
Wrong or bad connections.	Check for proper voltage connections and repair.
Low voltage, overloaded line, or wiring too small.	Reduce load. Increase size of wire.[1]

[1] These repairs should be made by an experienced electrician.

SOLENOIDS

A solenoid is simply an electromagnet that is arranged to pull or push something when power is applied to it. Figure 5-22 shows a simple solenoid. It consists of a winding of many turns on a hollow form. In the center of the form is an armature made of soft iron or of permanent magnet material. When the coil is energized its magnetic field will pull the armature into the coil. Often a spring is used to push the armature back out of the coil when power is removed.

Solenoids can be used in applications where only a small amount of motion is required. The solenoid valve described in the preceding chapter is a good example. The plunger of the valve doesn't have to be moved very far to open or close the valve. Another very common application of the solenoid is in an electric door lock. Here the coil is

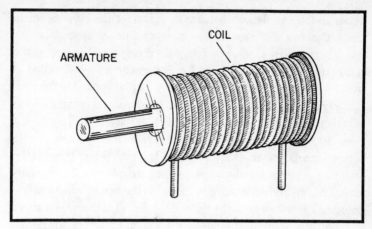

Fig. 5-22. Simple solenoid.

used to move the tongue of the lock so that it will lock or unlock a door or window.

In an application where a mechanism is used to open and close outside doors, the solenoid electric lock is very useful. A conventional lock isn't as useful because someone has to go to the door and unlock it before the control system can open it. With an electric lock, the door can be unlocked electrically whenever the control system tries to open it.

Construction of a solenoid device is usually quite a chore. It isn't easy to tell in advance just how much current should be used, nor how many turns of wire should make up the coil. It is usually much

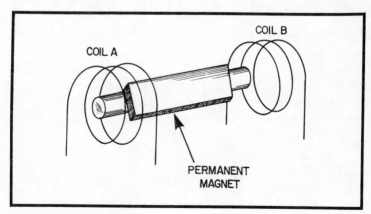

Fig. 5-23. Polarized solenoid.

better to take a solenoid that was designed for the purpose, or to convert an existing solenoid for use in a particular application.

There is no reason why a solenoid will only pull in one direction. If a permanent magnet is used for the armature, the direction of travel will depend on the polarity of the current applied to the coil.

Figure 5-23 shows a polarized solenoid that will provide a force in either direction. A short section of the armature is made of magnetic material such as soft iron. The rest of the armature is made of nonmagnetic material such as brass. The direction of travel of the armature depends on which coil is energized. When coil A is energized, the armature will move to the left in the figure; when coil B is energized, it will move to the right. If a solenoid of this type is used, the control system must be designed so that both coils will not be energized at the same time.

6

Electrical Control Devices

We stated earlier that the simplest control situation is when the device that we wish to control is electrically operated. A typical example might be a light or an electric motor. In such a case all we need is a control device that can be operated from a small low-voltage signal and will, in turn, control a much larger voltage or current. The most common electrical control devices are switches, relays, and solid-state devices.

Usually the control device simply turns a circuit on or off. In spite of its simplicity, the on-off control device is adequate for nearly every home control application. Of course, there are a few situations where an on-off control device is simply not adequate. One application that immediately comes to mind is where we want to control the intensity of a light. Here we need a device that will not only turn the light on and off but will in some way control the amount of light that it emits. Another obvious application where on-off control is inadequate is where we wish to control the speed of a motor. Here again we want a device that will not only turn the motor on but will also control how fast it turns.

In every case one of the principal objectives of the control device is to isolate the control signal from the power line. This isolation is necessary to avoid potential electric shock or fire hazards.

The most obvious device to use for remote control is a simple relay. Here a low-voltage signal applied to the coil of the relay controls a much larger voltage that is switched by the relay contacts. In selecting a control relay we must be sure that the contacts are capable of handling the voltage and current drawn by the device that we wish to control, and that the resistance of the coil is suitable for the control signal we wish to use. It is a good idea to use enclosed relays so that the contacts will not become contaminated with dirt or dust.

The principal limitation of the relay is that we must keep the coil energized continuously to hold the relay in one of its two positions. This imposes rather strict conditions on the power supply of the control system. It must be able to supply enough current to energize all of the relays in the system when necessary. Usually it is much better if we only need to send control signals when we wish to change something. The rest of the time there is no signal on the control line. A system of this type has rather simple power supply requirements and has practically no risk of electric shock or fire.

LATCHING RELAY

The latching relay is an ideal component for any remote or automatic control system. A latching relay is one that will latch in either of its two positions. The only time that we have to apply a signal is when we wish to change the relay from one state to another. This permits the control signals to be in the form of pulses that are only transmitted when we want to change the state of whatever we are controlling. The rest of the time there is no voltage or current on any of the control lines.

Latching relays are available commercially from electronics supply houses or from electrical lighting houses. They are used commercially for low-voltage control of ordinary home lighting. Unfortunately many commercially available latching relays of this type are quite expensive. It is better, if possible, to find them in surplus houses where they are usually available for much less.

If you can't buy a latching relay at a suitable price, you can always build your own. Figure 6-1 shows a sketch of a homemade latching relay. The heart of this relay is a reed switch that is operated magnetically. The contacts of the reed switch are normally open. When a magnet is brought near the switch, the contacts will close. In

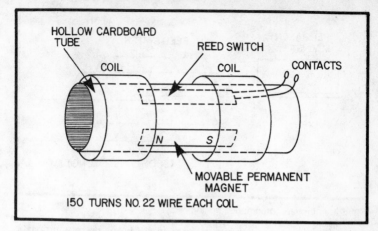

Fig. 6-1. Simple homemade latching relay.

the arrangement of Fig. 6-1 the reed switch is mounted in a small hollow cardboard tube. A small permanent magnet is also placed inside the tube and the ends of the tube are closed so that the magnet can't fall out. If the magnet is moved directly under the reed switch the contacts will close. When the magnet is moved a short distance away, still inside the hollow tube, the reed switch will open.

On the outside of the cardboard tube we have wound two coils of fine wire. The magnetic fields from either of these coils cause the permanent magnet to slide back and forward inside the tube, opening and closing the reed switch. To close the switch we simply energize the coil at the left of the figure long enough for the magnet to move under the reed switch. The current can then be removed. To open the switch we energize the coil at the right of the figure.

The coils are only energized long enough to slide the magnet to the proper position. The limitations of this homemade latching relay are that it must be mounted horizontally and it may operate erratically if it is subjected to excessive vibration.

AN IMPROVED HOMEMADE LATCHING RELAY

Although the homemade latching relay shown in Fig. 6-1 is fully adequate for many home applications, it has the limitation that the permanent magnet must move. This means that it must be mounted horizontally and may be subject to the influence of vibration.

If you are willing to do a little experimenting to find the correct dimensions and locations of the components, you can build a latching

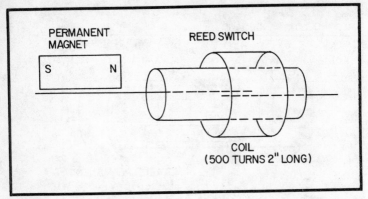

PERMANENT
MAGNET

| S | | N |

REED SWITCH

COIL
(500 TURNS 2" LONG)

Fig. 6-2. Improved homemade latching relay.

relay that is both simpler and more reliable. It has no moving parts except the reeds inside the reed switch and thus can be mounted in any position. It is no more subject to the influence of vibration than the reed switch itself.

The circuit of Fig. 6-2 works on the principle that it takes less magnetic force to hold a reed switch closed than it does to actually close it when it is open. The trick is to find the proper position of the bias magnet shown in the figure. Usually this magnet can be taped directly to one of the leads of the reed switch. The magnet must be positioned so that it will not cause the reed switch to close, but will keep it closed if another magnetic field is used to actually close it.

The best way to find the proper position of the bias magnet is to use two small magnets. Tape one of the magnets loosely to one of the leads of the reed switch. Connect an ohmmeter across the reed switch so that you can easily tell whether the contacts are open or closed. Then bring the other magnet close to the reed as shown in Fig. 6-3. Adjust the position of the bias magnet so that it will not cause the reed switch to close, but will keep it closed once the flux from the other magnet closes it. The relay should open and stay open when the other end of the magnet is brought close to it.

The coil is mounted so that its flux will add to or subtract from the flux from the bias magnet. It may be necessary to experiment a little to find the best position for the coil. Once a relay of this type has been designed experimentally, it should be possible to build others using the same type of coil, reed switch and magnet with no trouble. This latching relay will require a little more design time than the one shown in Fig. 6-1, but it will be much more reliable and easier to use.

The principal limitation of this type of relay is the current carrying capacity of its contacts. The reed shown in the figure can carry a current of about 1A if the load is noninductive.

To use a reed switch to control larger currents, it is advisable to use an SCR to actually control the load, and use the reed relay to turn the SCR on.

The latching relay of Fig. 6-2 only uses a single coil. It is latched by applying a current through the coil in one direction and is un-latched by running the current through the coil in the opposite direction. The relay of Fig. 6-1 uses two separate coils, one for latching and one for unlatching the relay. Most commercial latching relays also use the two-coil arrangement.

CONTROL SYSTEM FOR LATCHING RELAYS

The latching relay has the advantage that it can be operated from a low-voltage DC power supply. Inasmuch as current only flows when a relay is being latched or unlatched, the power supply requirement is very small. It doesn't hurt to overload a power supply for the second or so that is required to operate the relays.

The easiest way to latch and unlatch relays is to use a dual polarity power supply. The same arrangement can be used with two coil relays by adding two diodes to the relay as shown in Fig. 6-4. With this arrangement if the applied signal is positive with respect to ground, the current will flow through coil A, the latching coil. This will close the contacts of the relay. If the polarity of the applied signal is such that it is negative with respect to ground, the current will flow

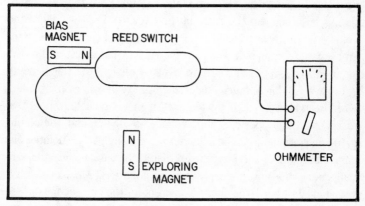

Fig. 6-3. Finding the right position for the bias magnet.

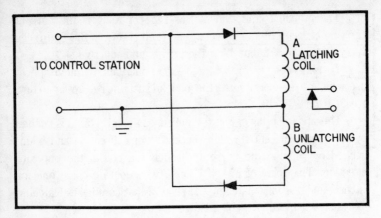

Fig. 6-4. Diode control for two coil latching relay.

through coil B, the unlatching coil. This will open the contacts of the relay.

The same power supply can be used with the single coil relay because the direction of current through the coil will change with the polarity of the applied signal as shown in Fig. 6-5. Thus it is possible with this power supply arrangement to use both single coil, and dual coil latching relays on the same system. In either case the relay will be opened by a signal of one polarity and closed by a signal of the opposite polarity. It doesn't matter which type of relay happens to be on any particular circuit.

If commercial latching relays are used in a system, design of the power supply is very simple. It only has to supply the rated operating voltage of the relays. If homemade relays are used, however, it is advisable to determine the optimum operating voltage. The way to do this is to connect the relay in the circuit of Fig. 6-6. Here the voltage is slowly increased until the relay latches. Then it is increased a little more. The coil of the relay is left connected to the supply for about 15 seconds to be sure that it will not heat up excessively. This is much longer than the relay will ever be energized in an actual control system, but it should be able to pass this test. Once the proper operating voltage has been found, the output of the power supply can be adjusted accordingly. As we pointed out earlier because of the intermittent nature of the latching relay arrangement, the power supply isn't at all critical. Most relays can withstand a rather large surge of current for the one second or less that it takes to perform the latching or unlatching operation.

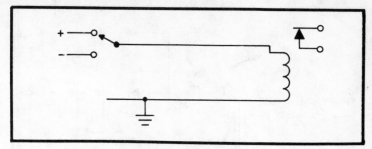

Fig. 6-5. Dual supply with single coil latching relay.

Figure 6-7 shows the schematic diagram of a dual polarity power supply suitable for operating a latching relay system. It actually uses two separate half-wave rectifiers—one to supply the positive voltage and the other to furnish the negative voltage. The filter capacitors are not really necessary, but without them the relays may chatter annoyingly when the state is being changed. The supply shown can be set for either a 6 or a 12 volt output. Usually this will take care of any homemade latching relay. If the power supply is only being used to control latching relays, a transformer rating of one amp is fully adequate. If the supply will also be used to energize other circuits such as indicators, the transformer should be rated accordingly.

The fuse in the primary part of the circuit is a very important part of the supply and shouldn't be left out. The supply of a control system is left energized at all times and should be protected against drawing excessive currents which might cause a fire. If the load is very heavy, it will be helpful to use a slow-blow type of fuse.

The power supply can be built right into the control box of the system, but this means that it will be necessary to run the power line

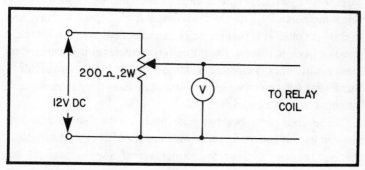

Fig. 6-6. Circuit for finding relay voltage.

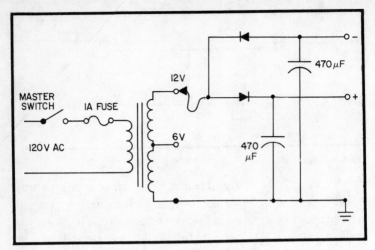

Fig. 6-7. Simple power supply for control system.

into the control box. This is satisfactory, if precautions are taken to avoid the possibility of an electric shock in the event of a component failure. In many cases it is advisable to mount the power supply in another location and run the operating voltages to the control box. In this way, the highest voltage in the control box will only be about 12V DC and the arrangement will be completely safe.

CONTROL SWITCHES

The control switching arrangement for a system using latching relays is extremely simple. Figure 6-8 shows a typical arrangement. There are three connections between the control box and the power supply—lines carrying +12V, −12V, and ground. One side of every control circuit is grounded. The switches are all momentary action push button switches. When switch S1 is closed, the positive supply line is momentarily connected to terminal A which goes to one of the control circuits. This positive pulse of current will cause the corresponding latching relay to latch, turning on whatever is connected to the circuit. When switch S2 is closed momentarily, the negative supply line will be connected to terminal A, unlatching the relay and turning the controlled device off.

The diagram shows two additional circuits, but as many as desired can be added. Because only impulses are used for control signals, there is no limit to the number of devices that can be controlled by this simple switching arrangement.

126

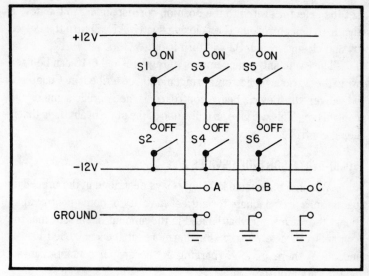

Fig. 6-8. Control switch wiring.

A very small control box may be made by using a selector switch as shown in Fig. 6-9. Here there is only one on switch and one off switch. Again they are momentary action push button switches.

Closing switch S1 momentarily will apply a positive pulse to the wiper of the selector switch. The circuit that this pulse actually goes to is determined by the position of the selector switch. Thus if the switch is set to position A, the pulse will go to the latching relay corresponding to device A, turning it on. To use the system, simply

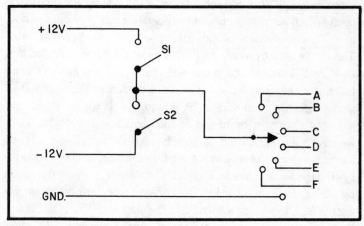

Fig. 6-9. Using a selector switch for control.

set the selector switch to the position corresponding to the device that is to be controlled. Then closing switch A1 will turn the device on and closing switch S2 will turn it off.

The switching arrangements of Figs. 6-9 and 6-10 can be used with the various indicator arrangements described in Chapter 2 whenever the device being controlled is not within sight of the control box. This will give an indication of the state of anything that is being controlled.

SOLID-STATE CONTROL DEVICES

Although the simple latching relays described in the preceding paragraphs are adequate for a great majority of home control functions, they are somewhat limited in the amount of current that the contacts can safely carry. Furthermore, if the controlled load is inductive, there may be sparking at the switch contacts causing deterioration of the contacts.

Increased current carrying capacity as well as freedom from contact problems can be obtained by using solid-state devices to switch the actual power.

Figure 6-10 shows the schematic diagram of a silicon controlled rectifier (SCR). This device has three electrodes; an anode, a cathode, and a gate. The anode and cathode are quite similiar to those in an ordinary diode. The difference is that no current will flow between the anode and the cathode of the SCR until a signal has been applied momentarily to the gate. Thus with the SCR connected as shown in Fig. 6-10 no current will flow. It will look like a switch that is open. When a signal is applied to the gate for a very brief period, about 5 millionths of a second (5 microseconds), the SCR will "fire." This can be done by momentarily closing switch S1. Current will now flow freely between the anode and the cathode as in an ordinary rectifier diode. The forward voltage drop across the SCR will be very small. Note that once an SCR is turned on we no longer have any control over it by means of the signals applied to the gate. Removing the signal from the gate has no effect at all on the anode current. In fact the only way we can turn the SCR off is to temporarily stop the anode current. This could be done by opening the circuit by means of switch S2 in Fig. 6-10 or by momentarily short circuiting the SCR by closing switch S3 in the diagram. When this is done the anode current will stop and the SCR will be reset to its original

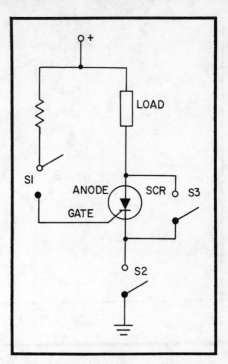

Fig. 6-10. Simple SCR circuit.

nonconducting condition. Note that it takes a much longer period of time to shut the SCR off then it does to turn it on. When we interrupt the anode current we must wait at least 50 microseconds before reapplying forward voltage to be sure that the SCR will not re-trigger.

The SCR has one rather unpleasant characteristic. That is it will turn on if the anode voltage is applied too suddenly. Thus sometimes an SCR will turn on when power is applied to the circuit even though no control signal is applied to the gate. This propensity of the SCR to turn on when voltage is applied is called the dv/dt effect. It can be minimized by the use of a "snubber" circuit. Details of snubber circuits are given in a later paragraph.

Figure 6-11 shows a different type of SCR. Here the gate is not used for triggering. Instead, the SCR is turned on by applying a beam of light to a photosensitive area behind a glass window in the case. The beam of light has exactly the same effect as a signal applied to the gate of a normal SCR. Because this type of SCR is light activated it is called a light activated SCR or simply LASCR. When the LASCR is dark no current will flow through it. As soon as light is applied to

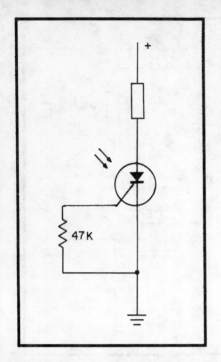

Fig. 6-11. Light activated SCR (LASCR).

the window it will turn on. As with a regular SCR, the only way we can turn it off is to momentarily interrupt the anode current by opening the circuit or by shorting the anode and cathode leads together. Note that removing the light has no effect at all on the anode current.

Both of the SCRs that we have described so far conduct current in one direction only. In this respect they are very similiar to the rectifiers used in ordinary power supplies. They are most suitable to use in DC circuits. If they are used in AC circuits they will rectify the current and will only deliver power during one half cycle of the applied voltage.

Figure 6-12 shows a device called a triac that is similiar to an SCR except that it conducts current in both directions. The triac can be turned on in either direction by a small gate current of either polarity. Because the triac conducts in both directions it never has any high reverse voltage as is sometimes found in rectifiers and SCRs. If the voltage applied in either direction should become too high the triac will simply be turned on. Most triacs are capable of conducting very high currents.

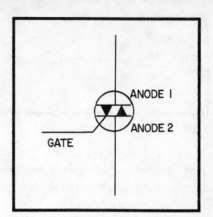

Fig. 6-12. The triac.

Figure 6-13 shows a typical triac circuit. Here when switch S1 is closed a small gate current will be supplied to the triac. It will thus turn on and supply power to the lamp in the circuit. Note that switch S1 must be kept closed for the triac to conduct. Otherwise the anode current will be interrupted every half cycle and the triac will stay in the off state.

Figure 6-14 shows a two-terminal device called a diac. It will not conduct current in either direction until the voltage across it reaches some breakdown voltage. Inasmuch as there are actually two diodes in the unit it can be turned on by an applied voltage of either polarity. Diacs are used to develop sharp pulses of current for turning on other semiconductor devices such as SCRs or triacs.

SNUBBER CIRCUITS

We stated earlier that an SCR will turn on if the anode voltage is applied too fast. Each SCR has a critical time factor. If the anode voltage is applied in a shorter period than this critical time the SCR will turn on regardless of whether or not a signal is applied to the gate.

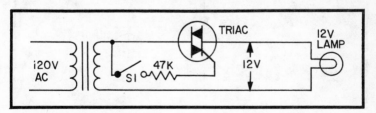

Fig. 6-13. Simple triac circuit.

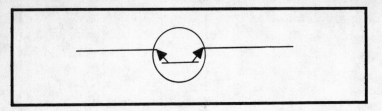

Fig. 6-14. The diac as switching diode.

Usually 60 Hz AC will not cause false triggering if the load is purely resistive. If the load is inductive, however, the voltage rise may be fast enough to cause false triggering. To avoid this false triggering, snubber circuits are used.

Figure 6-15 shows a typical snubber circuit. It consists of a resistor and a capacitor in series directly across the SCR. The design of snubber circuits is rather involved and will not be treated in detail here. In the circuits described in the book, the snubbers have been designed to give satisfactory performance. However, the reader will probably not use exactly the same load devices that the author used. For example, a relay or a motor that happens to be available to the reader may have more or less inductance than those used in the original design. If spurious triggering is encountered it can be cured by experimenting with the snubber circuit to determine proper component values.

RADIO AND TELEVISION INTERFERENCE

SCR type controls switch very rapidly and hence tend to generate high frequency energy. This high frequency energy can cause annoying radio and television interference. If the motor is only used

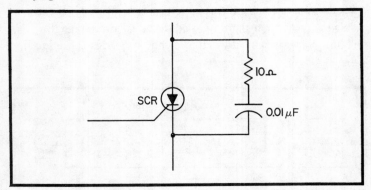

Fig. 6-15. SCR snubbing circuit.

occasionally for short periods to perform an operation such as opening and closing draperies the interference will probably not be very objectionable. If on the other hand an SCR system is used to control the speed of the motor which will be running for a substantial period of time the resulting radio and television interference may be severe. The first step necessary to minimize interference from an SCR control is to completely shield the control circuit. This will be all that is needed in some cases. In other cases it will be necessary to install filters in the leads running to and from the control system.

Figure 6-16 shows a typical radio interference filter. The capacitors are commercially available and the inductance can be made easily from available materials. The inductance in Fig. 6-16 is made by winding ordinary enameled wire on a ferrite core of the type used as an antenna on transistor radios. By carefully shielding the control circuit and installing filters in the leads interference can usually be reduced below the objectionable level.

TRIAC CONTROL OF AC LOADS

Figure 6-17 shows a circuit in which a triac is used to remotely switch AC circuits on and off. The gate of the triac is connected to the power line through transformer T1 at one end and through a variable resistor at the other end. The transformer T1 is an ordinary 120V to 6V transformer. The circuit is designed so that when the secondary of the transformer is open there will not be enough

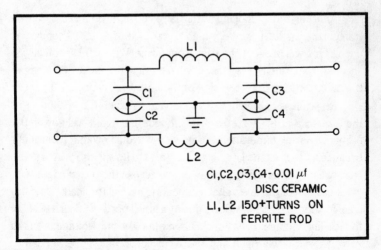

CI,C2,C3,C4 - 0.01 μf
DISC CERAMIC
LI, L2 I50+TURNS ON
FERRITE ROD

Fig. 6-16. Radio interference filter.

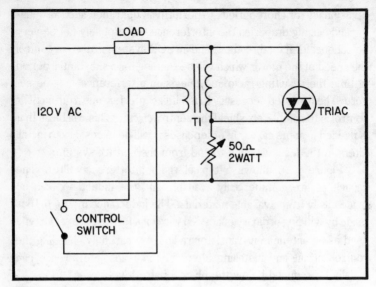

Fig. 6-17. Triac circuit for on-off control.

primary current to fire the triac. With some transformers, the leakage inductance of the transformer might cause enough primary current to fire the triac. Resistor R1 is variable and can be adjusted so that the triac will not fire when the secondary is open. When the secondary of transformer T1 is shorted by a switch at the control point a large amount of primary current will flow causing the triac to fire.

Another way of looking at this is to consider the primary of the transformer as having a very high resistance when its secondary is open. This will prevent the triac from firing. When the secondary is shorted, the primary resistance will decrease drastically connecting the gate of the triac to one side of the line, thus causing it to fire.

Regardless of which way you look at it, when the secondary of the transformer is shorted a large amount of current will flow in the primary. As any experimenter knows this would ordinarily burn out because as soon as the large current starts to flow in the primary, the triac will fire. This will drop the voltage across the transformer with practically all of the line voltage appearing across the load. Thus the transformer current only flows during a small portion of each cycle of the AC line voltage. A half cycle later when the line voltage drops to zero, the triac will open and the whole process will be repeated.

Thus we must keep the switch that is connected to the secondary of the transformer closed as long as we want power applied to the load.

This circuit is very practical for control of almost anything around the home. With the components shown in the figure it will control loads up to about 500 watts. About the only limitation of the arrangement is that a small pulsating current will flow in the control line all the time that the load is energized. This means that a momentary action switch cannot be used for control. The pulsating current might also induce hum in intercom lines if they are run in the same cable as the control lines.

A LATCHING RELAY WITH A TRIAC

When we discussed the latching relay earlier in this chapter, we mentioned that its principal limitation was that the current carrying capacity of the contacts, in many latching relays, is somewhat limited, precluding its use in controlling heavy loads. Figure 6-18 shows a latching relay used in conjunction with a triac to control heavy loads. The load current that this arrangement can control is limited only by the rating of the triac.

In this circuit the gate of the triac is normally open so it will not fire. When the latching relay is energized, the gate will be connected to one side of the line causing the triac to fire every half cycle applying power to the load. This circuit has all of the advantages of

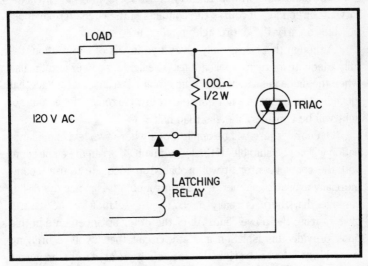

Fig. 6-18. Triac with latching relay.

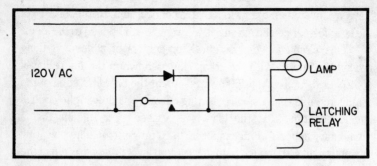

Fig. 6-19. Latching relay and diode for light dimming.

the latching relay without any of its limitations. Power is only applied to the circuit long enough to turn it on or off, so momentary action control switches can be used. The result is that there is no voltage or current on the control lines except when the state of the controlled device is being changed.

DIMMING OF LIGHTS

Figure 6-19 shows a circuit that will provide two levels of illumination. When full illumination is desired, the power line is connected directly to the lights. For reduced illumination, a half wave diode of the type that is normally used as a power supply rectifier is connected in series with the lighting circuit.

The diode must be able to carry the entire lighting current in the forward direction. Of course the contacts of the latching relay must be able to carry the entire lighting current.

Although this circuit only gives two levels of illumination it is fully adequate for many applications. The lighting power is cut in half when the diode is switched into the circuit. Because of the fact that incandescent lamps are not efficient at low currents, the amount of light will be considerably less than half.

Of course there will be applications where two levels of illumination will not be suitable. If the lights involved are in the same room with the control center, it will probably be adequate to use a commercially available dimmer at the control point. This has the disadvantage that there is really no complete isolation of the control system from the power line. All of the other arrangements in this book provide this isolation and any circuit that involves bringing something to the control box that is not fully isolated from the power line is definitely not recommended.

136

7

Tone Operated Systems

The control systems described in Chapter 6 that use low DC voltages as control signals are ideal for most home control applications. These systems do, however, have a couple of limitations. First of all, they require separate wires for each device that is to be controlled. This isn't much of a limitation in a small home, but in some applications it would be much simpler if all of the devices that are to be controlled could be connected across the same pair of wires.

Another limitation of the DC control system is that DC signals cannot conveniently be transmitted over telephone, carrier current, or radio links.

The tone operated systems described in this chapter have none of these limitations. As shown in Fig. 7-1, in a tone operated system, all of the devices that are to be controlled are connected across a single pair of wires from the control center of the system. The price that we have to pay for this nicety is that both the control box and the equipment that does the controlling will be more complicated.

At the control point of a tone-controlled system, separate audio tones are generated for each control function. The circuit that generates the tones is usually called a *tone encoder*.

Each of the controlled devices responds to a single audio tone and ignores all other frequencies that might be present on the control line. The circuit that responds to a particular tone is called a *tone decoder*. The output of the decoder is usually a DC signal.

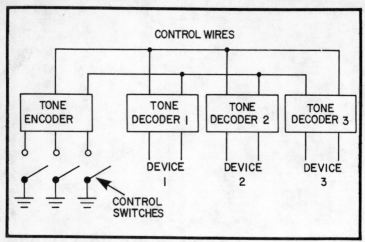

CONTROL WIRES

TONE ENCODER

TONE DECODER 1

TONE DECODER 2

TONE DECODER 3

DEVICE 1

DEVICE 2

DEVICE 3

CONTROL SWITCHES

Fig. 7-1. Tone operated control system.

It is easier to describe a tone-controlled system by starting with the decoder and working back toward the encoder.

DECODERS

There are two general ways of decoding audio tones. One is to use a filter that will pick out a particular tone. Usually active filters employing integrated circuits are used for this purpose. Another, simpler, approach is to use a phase-locked loop (PLL). In the circuits in this chapter we have chosen the phase-locked loop as a decoding device because of its simplicity and ease of adjustment. Before getting into how a decoder works we will take a look at the phase-locked loop and see what it is and how it works.

Figure 7-2 shows a block diagram of a phase-locked loop. It consists of a phase comparator, a low-pass filter, an amplifier, and a voltage-controlled oscillator, all contained in an integrated circuit.

In operation two signals are applied to the input of the phase comparator—one is the input signal and the other is a signal from the voltage-controlled oscillator. If the two signals were locked together, and were exactly 90 degrees out of phase, there would be no output from the phase comparator. If the two signals are not locked together there will be an output, called an error signal, which is amplified and applied to the voltage-controlled oscillator. The error signal will change the phase and if necessary the frequency of the signal from the voltage controlled oscillator until the output is locked

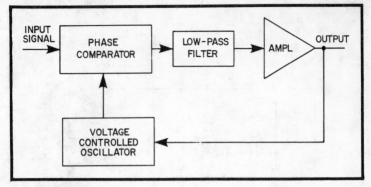

Fig. 7-2. Phase-locked loop.

to the input signal. The two signals will then be exactly 90 degrees out of phase. The low-pass filter is included to prevent the circuit, which is really a closed loop, from oscillating.

The Type 567 integrated circuit is a phase-locked loop designed particularly for frequency sensing or tone decoding. Like any other phase-locked loop, it has a controlled oscillator and a phase detector. In this particular circuit a second auxiliary, or quadrature, phase

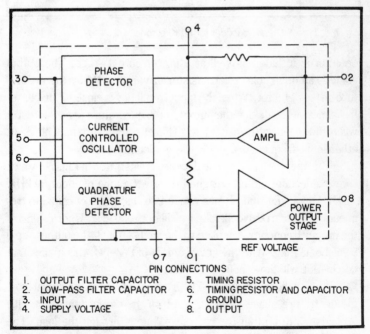

Fig. 7-3. Simplified block diagram of the Type 567 tone decoder.

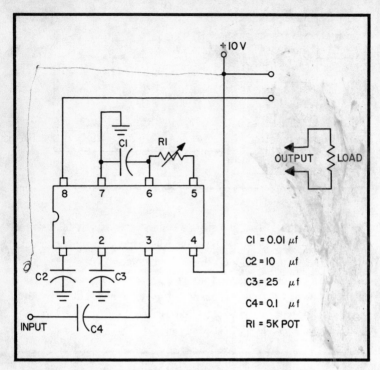

Fig. 7-4. Type 567 tone decoder for 50 KHz tone.

detector is included as well as a power output stage. The output stage is an open collector transistor driven directly by the quadrature phase detector. When the input frequency is close to the center frequency of the device the output of the quadrature phase detector will turn the output transistor on. This effectively grounds pin 8. At other input frequencies, pin 8 will be floating.

The operation of the tone decoder is illustrated in Fig. 7-4 which shows a decoder circuit designed to detect a tone of about 50 kHz. The values of R1 and C1 are chosen so that the free-running frequency of the current-controlled oscillator is 50 kHz. Until a signal of the proper frequency is applied to the input at pin 3, the output at pin 8 will be floating. With the arrangement of Fig. 7-4, this means that no current will flow though the load.

When a signal of about 50 kHz is applied to pin 3, a signal from the quatrature phase detector will turn the output stage on. This grounds pin 8, so that the current will flow through the load. The Type 567 will handle up to 100 mA of current.

The tone decoder is made immune to the effects of noise by adjusting the amount of time required for it to respond to a tone. This is accomplished by adjusting the bandwidth of the circuit. Wideband circuits respond rather quickly, whereas narrowband circuits are more sluggish in their response. This sluggish response prevents the circuit from being triggered by noise.

The center frequency of the Type 567 tone decoder is determined by the values of R1 and C1 which determine the free-running frequency of the current-controlled oscillator. The bandwidth of the circuit is determined by the value of C2, the applied signal voltage, and the center frequency.

The center frequency which can be set anywhere between 0.01 Hz and 500 kHz is determined by the values of C1 and R1. The value for R1 is given by

$$R1 = 1/fo\ C1$$

where fo is the center frequency. Before we can use this equation we must know the value of C1. This is chosen so that R1 will always be greater than 2000 ohms and less than 20,000 ohms. Table 7-1 shows values of C1 that can be used to keep R1 within this range for various tone frequencies.

The best way to handle the value of R1 is to use a 2000 ohm fixed resistor and a 15,000 ohm potentiometer in series as shown in Fig. 7-5. Then the proper value of C1 can be selected from the table in the figure to get the desired frequency range.

The values of C2 and C3 can usually be 10 μF and 20 μF respectively. If the circuit response is too slow these capacitors can be reduced; if the circuit is vulnerable to noise they can be increased.

Table 7-1. Values of Capacitor C-1.

FREQ. RANGE (Hz)		C1
LOW	HIGH	(μF)
10	100	4.7
100	1000	0.47
1000	10,000	0.047
10,000	100,000	0.0047

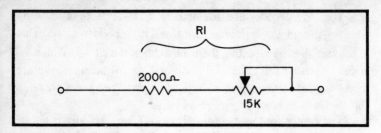

Fig. 7-5. Arrangement for R1 in Fig. 7-4.

TOUCH-TONE SYSTEM

One of the most useful tone-operated systems is the Touch-Tone dialing system used by the telephone company. In this system each digit is represented by a combination of two tones. The frequencies of these tones are chosen to avoid harmonics that might cause false triggering, and to avoid the fundamental and harmonics of the power line frequency.

The use of two separate tones to represent each digit, or control signal, has many advantages. Inasmuch as both tones must be present for a valid control signal to exist, the system is practically immune to spurious triggering. The system also has the advantage that a standard Touch-Tone telephone set can be used to generate the tones. Figure 7-6 shows the frequencies used in this dialing system. The system can be used to represent each of the numbers from 0 through 9 plus two additional symbols, usually * and #. It also can be used to represent letters of the alphabet as shown in the figure.

Whenever one of the buttons is pressed, two frequencies, one from the low group frequencies and one from the high group frequencies is applied to the line. For example, when the "5" button is pressed tones of 770 Hz and 1336 Hz are generated.

TOUCH-TONE ENCODER

Building a highly accurate Touch-Tone encoder is very easy because integrated circuits are available that will perform almost all of the functions. A typical circuit of this type is the Motorola Type MC 14410 tone encoder. A block diagram is shown in fig. 7-7a, and the pin connections are given in Fig. 7-7b. The various tones are derived from a 1 MHz signal generated by a crystal oscillator. This frequency is divided down digitally to provide the required tone frequencies. The inputs to the circuits are simply switch closures from a regular telephone Touch-Tone dialing pad.

This circuit has just about all of the features that might be desired in a tone generator. Diode protection is provided on all inputs. The noise immunity of the circuit is such that it will not respond to noise pulses up to 45% of the supply voltage. A multiple key lockout feature is built right into the integrated circuit so that no significant signal will be generated when two keys are pressed at the same time.

This circuit is extremely interesting from the point of view that it actually synthesizes a sine wave from digital signals. The digital signals are added together in such a way as to make a step type of approximation to a sine wave as shown in Fig. 7-8. When an external filter is added to the circuit all harmonics are at least 30 dB below the fundamental.

The circuit uses CMOS technology and so it draws very little power and has a high noise immunity. In addition to the frequencies shown in Fig. 7-6, the encoder will also generate a frequency of 1633 Hz. This permits using a 4 × 4, rather than the standard 4 × 3

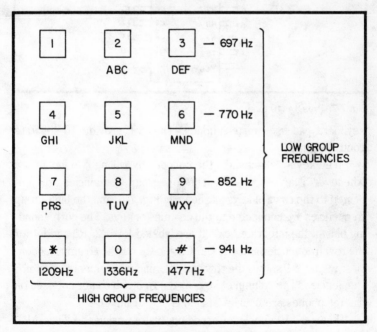

Fig. 7-6. Frequencies used in the Touch-Tone dialing system.

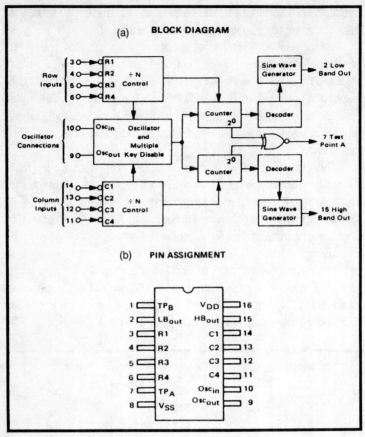

Fig. 7-7. Type 14410 tone encoder.

keyboard pad used in telephones. This will provide 16 separate control signals.

The keyboard activates the encoder by pulling two inputs low whenever a key is pressed. The low-group frequencies are connected to the rows of keys as shown in Fig. 7-6, and the high-group frequencies are connected to the columns of keys. The corresponding pins of the integrated circuit are labeled R1, R2, R3 and R4 for the row frequencies and C1, C2, C3 and C4 for the column frequencies. Figure 7-9 shows the row and column arrangements using all 8 frequencies. The column of keys at the right which are not used on dial telephones are usually labeled A, B, C and D.

Figure 7-10 shows an encoding system using the Type MC-14410 tone encoder. The high-group and low-group frequencies are

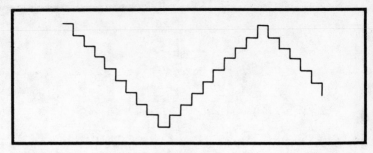

Fig. 7-8. Step approximation of a sine wave.

combined in a resistor network. If a 4 × 4 keyboard is not available a standard 4 × 3 keyboard may be used by simply ignoring the connections shown to the right hand column of keys. The supply voltage required by this circuit is 5V making it fully compatible with TTL logic.

Figure 7-11a shows a block diagram of another tone encoder suitable for use with a telephone keyboard pad and Fig. 7-11b shows its pin connections. This is the Mostek Type MK 5085. This IC has the advantage that it doesn't require a 1 MHz crystal, but can operate from a low-cost 3.58 MHz "color subcarrier" crystal of the

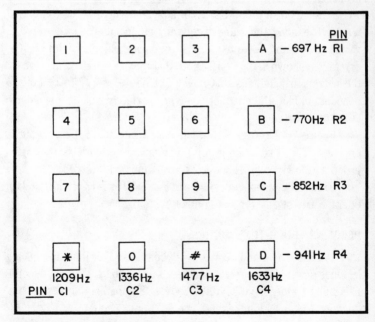

Fig. 7-9. 16 key tone arrangement.

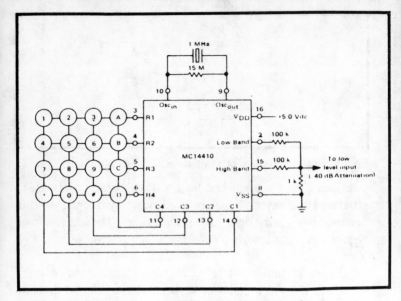

Fig. 7-10. Encoding system using the Type 14410 tone encoder.

type used in color TV sets. Tones are generated by connecting the proper R or C pins to the positive supply voltage.

When using a 3.58 MHz color subcarrier crystal the tones will not have exactly the same frequency as specified earlier for the various rows and columns. The frequencies obtained with such a crystal are shown in Fig. 7-12. Note that they are so close in value to the correct frequencies that it will not be necessary to re-tune a decoder. In fact, the encoder will work perfectly well on a standard telephone system.

Figure 7-13 shows a tone encoder using the Type MC 5085 encoder. There is no need for a combining network as the high-group and low-group frequencies are combined internally.

Either of the encoders shown in Fig. 7-10 or 7-13 can be used to operate almost any type of control system.

DECODING TOUCH-TONE SIGNALS

Because the standard 4 × 3 keyboard used for tone dialing can produce seven different frequencies, a full decoder that can decode all 12 digits must have seven decoders. Figure 7-14 shows such a decoder. It consists of seven Type 567 tone decoders and 12 NOR gates. The NOR gates can be three quad NOR gates so only three

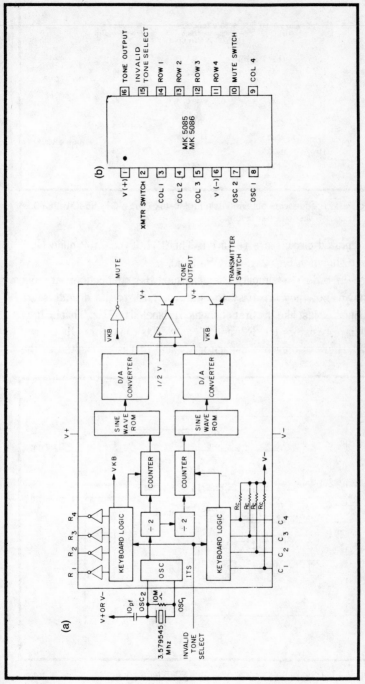

Fig. 7-11. Block diagram and pin connections for Mostek Type 5085 dual tone generator.

147

		Standard DTMF (Hz)	5085/6 Tone Output Frequency Using 3.579545 MHz Crystal	% Deviation From Standard	
ROW	f_1	697	699.1	+.31	
	f_2	770	766.2	−.49	LOW GROUP
	f_3	852	1 847.4	−.54	
	f_4	941	948.0	+.74	
COL	f_5	1209	1215.9	+.57	
	f_6	1336	1331.7	−.32	HIGH GROUP
	f_7	1477	1471.9	−.35	
	f_8	1633	1645.0	+.73	

Fig. 7-12. Frequencies obtained using a Mostek Type MC 5085 IC with a 3.58 MHz. color TV subcarrier crystal.

integrated circuits are required. The TTL Type 7402 quad NOR gate shown in Fig. 7-15 may be used.

In order to simplify the diagram, the connections are not shown, but they are easy to figure out. All of the decoders are connected just like the first one which is labeled 697 Hz. That is, they all have resistors R1, R2, and R3 as well as capacitors C1, C2 and C3. There is only one capacitor C4 and that is connected to the

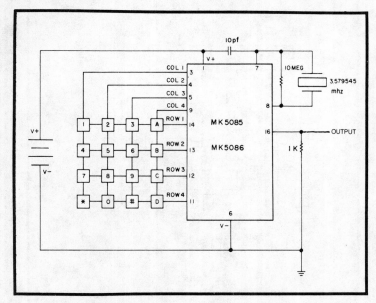

Fig. 7-13. Tone generator using Mostek 5085 tone generator.

junction of the connections from all of the R2s from the seven decoders.

The outputs from pin 8 of each of the decoders are connected to the inputs of the NOR gates. Again, the connections are easy to

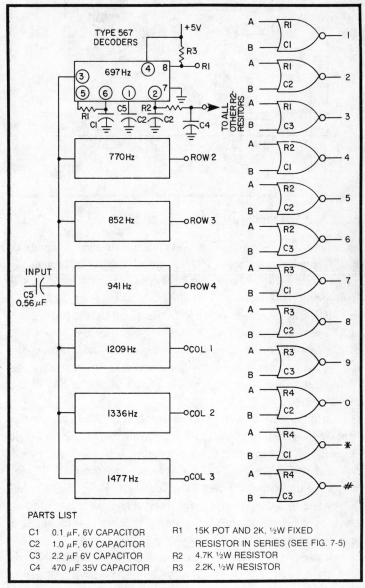

Fig. 7-14. Full 12 digit dual tone decoder.

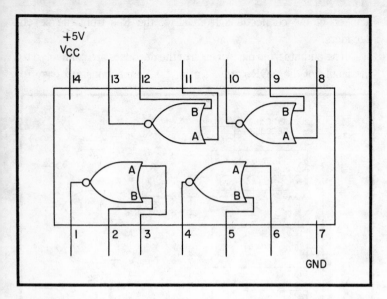

Fig. 7-15. Pin connections, Type 7402 quad NOR gate.

figure out and are shown in Table 7-1. For convenience, the outputs of the decoders are labeled R1 through R4 and C1 through C3 corresponding to the rows and columns on a standard 4 × 3 keyboard.

NOR gates are used because NORing high signals is the same as ANDing low signals. That is, the only time that the output of a NOR gate will be low is when both of its inputs are low. The decoder outputs go low when they receive the proper tone, so we can use the NOR gates to tell when the outputs of any particular two of the decoders are low.

To understand how the various decoders are connected to the inputs of the gates, refer back to Fig. 7-6. Here we see that the digit "1" is in the first row and the first column. Therefore we should expect that the two inputs of the gate corresponding to "1" will connect to the outputs of decoders R1 and C1. Referring to Table 7-2 we see that this is indeed true. The connections to the other gates can be figures out the same way, but it isn't necessary because all of the connections are shown in Table 7-2.

Before the circuit can be placed into operation all of the decoders must be tuned. Even though each of the decoders will be tuned to respond to a single tone, a two-tone signal can be used at the input for tuning. To find out when the decoder responds, connect a voltmeter from pin 8 of each decoder to ground. Normally this pin will be high. When the decoder responds to a tone, the voltage at pin 8 will drop to very nearly zero. To get the proper frequency for tuning, simply press a key in the row or column corresponding to the particular decoder. Thus decoder R1 can be tuned by pressing any key in row 1, that is, any one of keys 1, 2, or 3. The other decoders can be tuned in the same way.

The circuit of Fig. 7-14 is ideal at a main terminal in a control system. For example, it could be used at the receiver of a radio comtrol system. However in a system where wires are used to carry the control signals to the various devices that are to be controlled, it will rarely be necessary to decode all twelve possible digits at any one location.

At some points a single decoder may be all that is required. At other points it may be necessary to decode a few more digits. In such cases, a part of the system of Fig. 7-14 can be used. It is only necessary to use enough decoders and gates to detect the tone combinations that will have some effect at a particular point in the system.

Figure 7-16 shows a circuit that can be arranged to decode either simultaneous or sequential tones. If the two capacitors C3 in the diagram are made the same value, say $2.2\mu F$, both tones must be present before the output of the NOR gate will go to the high

Table 7-2. Connections for Fig. 7-14.

DECODER OUTPUT	GATE INPUTS
R1	1A, 2A, 3A
R2	4A, 5A, 6A
R3	7A, 8A, 9A
R4	*A, 0A, #A
C1	1B, 4B, 7B, *B
C2	2B, 5B, 8B, 0B
C3	3B, 6B, 9B, #B

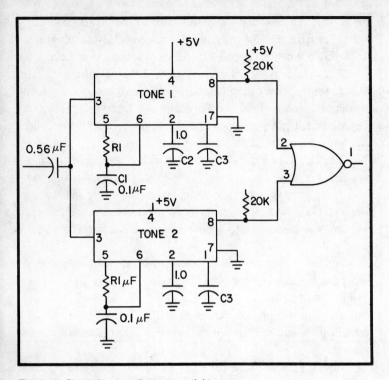

Fig. 7-16. Circuit for decoding sequential tones.

state. Thus we need two simultaneous tones to get an output. If, however, we make the capacitor C3 in the tone 1 decoder very large, say 470 μF, then its output will remain low until the second tone is transmitted. Note that the output won't go high if tone 2 is transmitted first.

Several of the circuits like that shown in Fig. 7-16 can be used in various parts of a control system. The result is that many rather elaborate functions can be controlled by a simple two-tone arrangement.

LATCHING

So far all of the circuits we have discussed will have a high output only when a tone is being transmitted. This is adequate for control functions that are intermittent in nature, for example, the arrangement would be fine for moving something like a door or drapery or energizing the push-to-talk relay of an intercom system. It would not be satisfactory for handling a continuous function such as

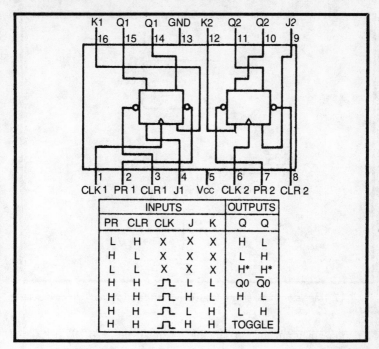

INPUTS					OUTPUTS	
PR	CLR	CLK	J	K	Q	Q̄
L	H	X	X	X	H	L
H	L	X	X	X	L	H
L	L	X	X	X	H*	H*
H	H	⊓	L	L	Q0	Q̄0
H	H	⊓	H	L	H	L
H	H	⊓	L	H	L	H
H	H	⊓	H	H	TOGGLE	

Fig. 7-17. Type 7476 J-K flip-flop.

turning on a light. It wouldn't be much of a remote control system if you had to hold a push button depressed to keep a light turned on. In such applications we need a latching arrangement.

One of the handiest circuits that can be used for latching the output of a tone decoder is the J-K flip-flop. Figure 7-17 shows a

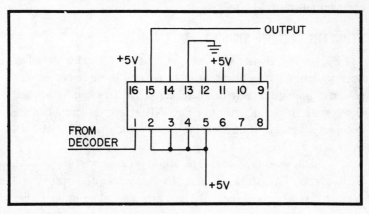

Fig. 7-18. J-K flip-flop used as a toggle.

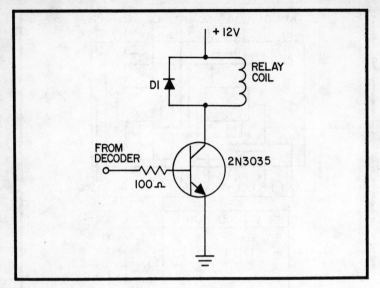

Fig. 7-19. Transistor amplifier for decoder. D1 can be an ordinary silicon rectifier diode.

functional diagram and a truth table for the Type 7476 TTL J-K flip-flop. This circuit can be connected to perform any complex logic operation on the outputs of decoders. One of the handiest arrangements for use in a control system is the toggle arrangement shown in Fig. 7-18. Here the flip-flop acts as a toggle. One tone burst will make the output high and the next burst will make it low. This means that the same tone or combination of tones can be used to turn the same device on and off. Thus it will save a great deal of hardware in a complicated control system.

USING THE DECODED SIGNALS

So far the signals that we get from the various decoding schemes are either +5V or zero. In general there is not enough power to operate a relay or a latching relay. This can be handled simply with the arrangement of Fig. 7-19. Here the output of the decoder is applied to the base of a transistor that will switch on current of a relay.

Probably one of the best applications is to energize the coil of a latching relay. When this is done, the tone operated system will operate just like the simple DC systems described in an earlier chapter.

8

Light Beam, Ultrasonic, and Audible Tone Systems

In all of the control systems described thus far, the control signals are carried from the control point of the system to the various devices that are to be controlled by means of wires. Indeed, this is usually the best and most reliable way to design a control system. There are instances, however, where it is impractical to use wires to connect the various devices of a system. For example, the application may call for a portable control box that can be carried around without any wires trailing from it.

One approach to the problem is to use radio control, as described in the final chapter of this book. Radio control presents unique problems, however, and it is usually easier to use some other control technique.

In this chapter we will consider the use of systems that use a light beam, an ultrasonic signal, or an audible tone to carry the control signal from the control point to the device being controlled. Any of these techniques eliminates the need for wires to carry the control signals and makes practical the use of portable control boxes. Furthermore, they have none of the disadvantages of radio control systems.

TV REMOTE CONTROL SYSTEMS

One very easy way to get a remote control system that doesn't use any wires is to salvage the remote control system from a TV set

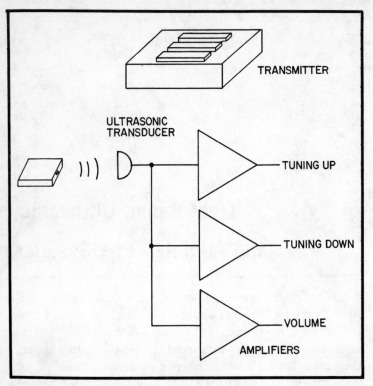

Fig. 8-1. Elements of a TV remote control system.

that is being discarded. Figure 8-1 shows the elements of a common TV remote control system. The remote control functions are usually:

Tuning Up
Tuning Down
ON OFF and Volume

The transmitter is often a mechanical arrangement that generates an ultrasonic tone burst by mechanically striking a bar of metal. The receiver is an ultrasonic transducer that produces an electric signal when an ultrasonic tone is received. The volume is usually controlled by a stepping switch. The contacts of this switch can be used directly to operate some controlled device.

The tuning control usually steps some sort of electric motor forward and backward to move the tuning turret. There is usually a snap action switch and cam arrangement that can be set to stop the tuner on any of the channels that are used in the particular area. This switch can be used as a pulse generator to generate electrical pulses to select a controlled device.

Figure 8-2 shows a TV control system that has four levels of volume as well as upward and downward tuning. This system was modified so that each of the volume positions selected a different control line. After the line is selected, the up-tuning signal will send a turn on signal down the line and the down tuning signal will transmit a turn off signal. The system can be used with the latching relays described in an earlier chapter. It is true that only four devices can be controlled with this particular arrangement, but it usually isn't necessary to control everything connected to a system from a portable control box. Often one or two devices are all that need be controlled. The rest can be controlled from the main control box.

The is no reason why the system has to be limited to the simple modification shown. A remote control receiver can be used to drive a stepping switch that will control any number of devices. The only problem will be constructing the interface between the salvaged unit from the TV set and the stepping switch. The details depend on the circuitry of the particular TV set from which the unit was salvaged.

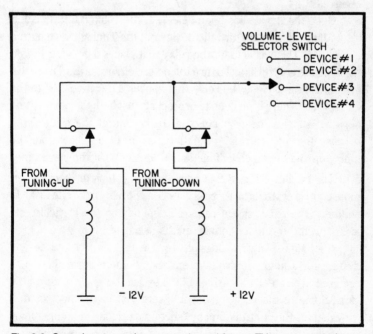

Fig. 8-2. Control system using parts salvaged from a TV remote control.

Because there are so many different circuits in use the details of the modification are beyond the scope of this book.

Figure 8-2 shows a partial diagram of a remote control system that was made from parts salvaged from a TV remote control system. The receiving unit taken from the TV set had three outputs. One was a stepping switch that switched resistance into the circuit to control the volume of the receiver. The resistors were taken from this unit and only the contacts used in the control system. The coil was left in the circuit so that the switch would advance when the volume button on the control box was pushed. The other two outputs were signals of about 9V that told the tuning motor to move one way or the other. Relays were connected to these outputs and used to determine the polarity of the control signals.

Looking at Fig. 8-2, the first step is to press the volume button on the remote control box until the volume level selector switch is in the proper position. The switch has four positions, each of which corresponds to a different device that is to be controlled. At the device, a latching relay is used that will turn the device on when a positive signal is received and turn it off when a negative signal is received.

Once the device has been selected, the volume button is left alone. If it is desired to turn the device on, the Tuning Down button is pressed. This will actuate the relay that places a positive 12 volt signal on the control line. To turn the device off, press the Tuning Up button. This will close the relay that applies a negative 12V to the control line. There is no need to protect against both the positive and negative supplies being connected to the line at the same time, because the original remote control system didn't allow both the tuning up and tuning down functions to be active at the same time.

The circuit of Fig. 8-2 is necessarily incomplete because remote control systems in TV sets vary considerably from one make to another. The exact circuit required depends on the particular set from which the remote control unit is salvaged.

It is almost always possible to use the TV remote control system for a home control system. Sometimes the modifications required are rather extensive, but they can be made with a little care. Some of the earlier color TV sets that had remote control provided for the control of many functions including not only tuning and volume, but also color, tint, and saturation. A system of this type would permit controlling many different devices without adding much to the system.

AUDIBLE TONES

A very simple, and yet very practical, control system uses audible tones to carry control signals from a control box to a central control point in the system. Of course, if control signals were sent at very frequent intervals, the audible tones could be annoying, but in most systems control signals are sent at rather infrequent intervals and the tones are not particularly unpleasant. The system has the advantage that the presence of the tones will indicate that the control box is working properly. This isn't particularly obvious in the case of radio or ultrasonic control systems.

Figure 8-3 shows the arrangement. The control box of the system is a tone generator and audio amplifier that drives a small speaker. The receiver is just like an intercom system. It has a permanent magnetic speaker that acts as a microphone and an audio amplifier. The output of the amplifier is fed to a tone decoder as in a tone operated system. In fact all of the arrangements described in Chapter 7 can also be used with audible tones.

To keep the sound output of the control box at a comfortable level, it is advisable to have a separate receiver in each room in which the control box may be operated. If the room is very large, two receivers can be used, one at either end of the room.

Of course, the receivers of this system will pick up all of the ambient noise and conversation in the rooms where they are located. If the two-tone system is used, this isn't a problem unless the

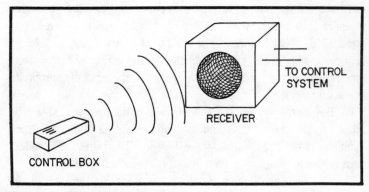

Fig. 8-3. Audible tone control system.

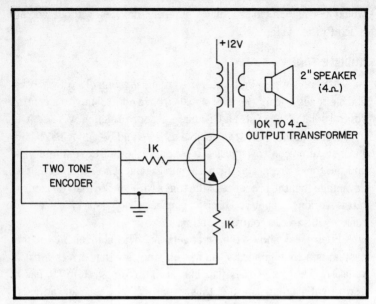

Fig. 8-4. Transmitter.

ambient noise level is very high, in which case it would probably be better to use some other type of system.

The control box of the audible tone system is shown in Fig. 8-4. The tone generator can be one of the tone generators described in Chapter 7, Fig. 7-14. The output of the tone generator is fed to a power amplifier that drives a 2-inch permanent magnet speaker. The speaker is mounted at one end of the control box so that it can be pointed at a receiver when it is being used.

Figure 8-5 shows the diagram of the receiver. It consists of a two stage intercom amplifier that is connected to a line through a regular output transformer. The output transformer is used so that unshielded speaker wire can be used to carry the signal back to the control center of the system. At the control center of the system, the lines from the receiver are connected to the input of appropriate tone decoders.

Before the system is installed, each of the decoders should be tuned to the proper frequency by connecting the output of a tone encoder directly to the input of the decoder. Then after the decoders are properly tuned, the system can be installed.

The only problem in installation is adjusting the gain of the amplifiers in the receivers. This is done by temporarily installing a

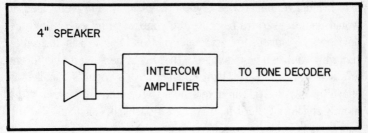

4" SPEAKER

INTERCOM
AMPLIFIER

TO TONE DECODER

Fig. 8-5. Receiver.

decoder at the receiver as shown in Fig. 8-6. The control box is then brought to the most remote point in the room in which it might be operated. The key of the encoder that corresponds to the decoder being used is then pressed. If the decoder doesn't produce an output, the gain of the amplifier must be increased. In most homes this is a simple adjustment and can be made with little difficulty.

LIGHT BEAM CONTROL

A light beam is a convenient way of transmitting control signals for simple on-off functions. Of course, it is possible to use a modulated light beam to provide complete control, but such systems tend to be hard to install and get working. For this reason we will not consider them here. We will, however, look at some circuits that can be used to send an on-off control signal where it is inconvenient to use any other type of transmission.

One place where a light beam type of control is handy is where the device that is being controlled might be moved around occasion-

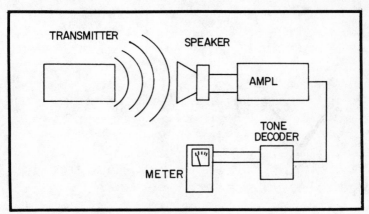

TRANSMITTER SPEAKER

AMPL

TONE
DECODER

METER

Fig. 8-6. Setup for tuning tone decoders.

ally. If a conventional wired type of control system were used, it would be necessary to re-route the control wiring every time the device were moved. For example, suppose that it was desired to remotely control a ventilating fan that might be moved to different locations in a home. If a light beam control system is installed as shown in Fig. 8-7, it will be possible to turn the fan on and off by merely shining a light on it. An ordinary flashlight can be used as the source of the control signal.

Several circuits may be used to detect the presence of a light beam. Usually the only problem in such a system is protecting the light sensitive device such as a phototransistor from getting too much ambient light. The easiest way to do this is to shield the device, leaving one area open for receiving the light beam.

Figure 8-8 shows a light detector circuit using a phototransistor. The phototransistor is shielded from ambient light and furthermore it is capacitively coupled to the rest of the circuit. This capacitive coupling will keep the circuit from responding to slow changes in the ambient light level. For example, it will not be affected by the gradual change from darkness to daylight.

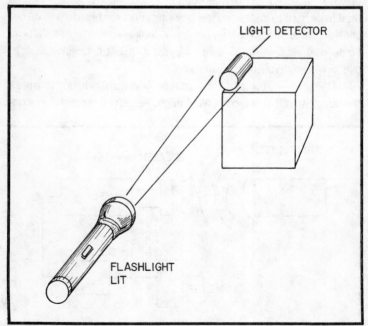

LIGHT DETECTOR

FLASHLIGHT
LIT

Fig. 8-7. Light beam control system.

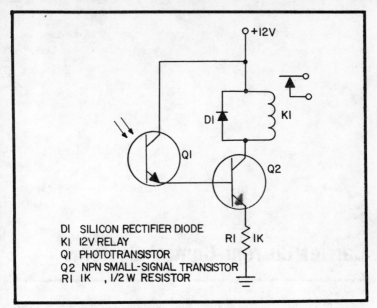

Fig. 8-8. Light sensitive receiver.

The output of the circuit of Fig. 8-8 can be connected to any of the electric control devices earlier. It may be used to operate a latching relay, or to operate a triac circuit.

The principal limitation of light beam control systems is that unless rather elaborate circuits are used, they will respond to sudden changes in light level. Thus, if at night the lights were turned on in a dark room, the photoelectric system would respond to the sudden change in light level. Of course, circuits can be designed to minimize this influence, but there are plenty of other ways of designing a control system and it is rarely worthwhile to use elaborate arrangements that are hard to get working properly.

9

Carrier Current Control Systems

When planning a control system, the question naturally arises as to why it is necessary to install control wires when the house is already wired. Why can't the power line be used to carry the control signal to the various parts of the home?

A system that uses the same wires that normally carry power to also carry control signals is usually called a *carrier current system*. Figure 9-1 shows the arrangement. Here the control signals are modulated onto an rf carrier which is fed to the power line. At the control point a receiver and demodulator is used to recover the control signal.

Superficially, the carrier current system looks like the best approach to the control problem. The control signals will be present at any point in the home where there is a power line which should include the location of almost anything that we might wish to control. There is a price that we must pay, however.

Household wiring was not designed to carry rf signals. Every device that might happen to be connected to the line, such as a lamp or an appliance, will short out some of the rf energy. Another limitation of the carrier current system is that a receiver is required at every control point, and if there is a sensor at the point, a transmitter will also be required.

For these reasons, the carrier current system isn't the ideal approach to a control system. It may, however, be used to advan-

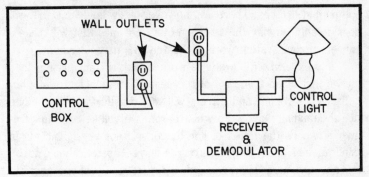

Fig. 9-1. Carrier current control system.

tage to get control signals to locations where installing control wiring would be very difficult. Probably the best compromise is to use regular control wiring where it is practical and use carrier current for the special applications where separate control wires are not practical.

THE PROBLEM

Figure 9-2 shows a simplified diagram of a home wiring system. The first thing that we notice is that there are two separate circuits. The power line entering the home consists of three wires. The voltage is about 120V between either of the two line wires and ground and about 240V between the two line wires. One of the circuits in the home is between one of the line wires and ground and the other circuit is between the other line and ground. In a carrier current system it may be necessary not only to send signals along the same circuit, but also from one circuit to the other.

The problem is complicated by the fact that not only are many devices such as lamps and appliances connected across the line, but we have no way of knowing which of these devices will be actually connected at the time that we wish to send a control signal.

Many of these seemingly impossible problems can be overcome by selecting the proper frequency for the carrier current. When this fails, it is possible to condition the wiring so that it will carry the control signals.

SELECTING THE FREQUENCY

Because of the fact that no home wiring situations are exactly the same, it is impossible to specify a single frequency that will be optimum in all cases. There are, however, a few guidelines that will considerably simplify the selection of a frequency that will make a carrier current system work adequately in a particular home.

Any experimenter who has worked with rf transmission lines knows that when a line is not terminated in its characteristic impedance the voltage and current will vary drastically along the line due to standing waves. For this reason, we will select a low frequency where the effect of line length will not be significant. In general, the longest path in the carrier current system should not be more than about 1/10 wavelength at the frequency of operation. In this case, the house wiring will look electrically like a network consisting of capacitors, inductances and resistors connected together in some unpredictable way. Bad as this is, it isn't as bad as when the wiring looks like a rather long transmission line.

In light of these considerations, most carrier current systems work at frequencies below the standard broadcast band. Frequencies between 60 and 180 kHz are common. Some systems use frequencies of 500 kHz and above, but these are very apt to cause radio interference.

In the description of systems that follows, we have selected a carrier frequency of 100 kHz, but there is nothing iron clad about this. The user may well find that signals of a different frequency will behave better in a given installation.

TONE FREQUENCIES

Fully as important as the choice of the carrier frequency is the judicious choice of tone frequencies that will be modulated onto the carrier. The cardinal rule in selecting tone frequencies is to avoid the 60 Hz power line frequency and its harmonics. Not only can the 60 Hz get into the transmitter and receiver and cause problems, but the presence of nonlinear devices connected to the power line can also generate harmonics. It isn't at all uncommon to find a substantial third harmonic—180 Hz—on the power line.

When the number of items being controlled justifies it, an ideal choice of tone frequencies is the set of dual tone frequencies described in Chapter 7 for the Touch-Tone dialing system. These frequencies were chosen after a great deal of study so that they would not coincide with the power line frequency or its harmonics. Furthermore, the fact that each digit is represented by two simul-

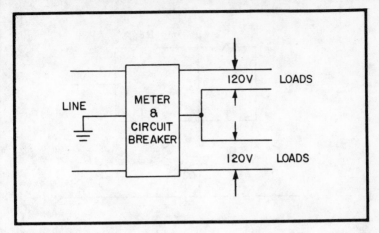

Fig. 9-2. Simplified household wiring system.

taneous tones means that even if a spurious signal coincided with one of the tones it would not cause trouble because both tones have to be present for the system to operate.

LOOKING AT THE POWER LINE

In order to get a carrier-current system working, we must know what types of spurious signals are present on the power line and we must have a way of tracing the carrier throughout the building in which the system is installed.

Measuring either spurious signals or the carrier itself in the presence of the 120V 60-Hz power is anything but easy. The 60 Hz voltage is so large that most conventional instruments will not provide much information. Even on an oscilloscope, the power line voltage is so large that it is nearly impossible to see anything else.

One of the handiest things to have when making the measurements required to get a carrier current system working is the Parallel-T or Twin-T filter shown in Fig. 9-3. This type of network will produce infinite attenuation at a single frequency while passing other frequencies with little attenuation. It is something like a bridge that is balanced at a single frequency, thus reducing the output to zero at this one frequency.

To get infinite attenuation at a single frequency, capacitor C2 is made twice the value of the two capacitors C1, and the two resistors R2 are each made twice the value of R1. The values of these

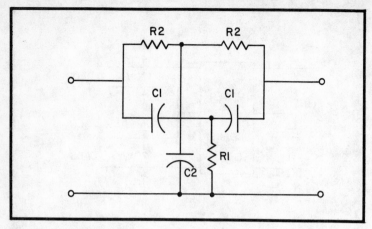

Fig. 9-3. Twin-T network.

components are then chosen from the formulas:

$$R1 = \frac{1}{2\pi C2f}$$

$$R2 = 2R1$$

$$C1 = \frac{1}{2\pi R2f}$$

$$C2 = 2C1$$

where f is the frequency that is to be rejected by the network.

Regardless of which of these equations we work with, we have not enough information to really pin the design. The values of the resistors depend on the values of the capacitors and the values of the capacitors depend on the values of the resistors.

The circuit shown in Fig. 9-4 was designed around some 9 μF capacitors that happened to be available. These capacitors are frequently available through surplus houses. If they aren't available, 1.0 μF capacitors which are widely available can be connected in parallel to get the desired capacitance. Using this value of capacitance, the value of resistor R1 turned out to be 147 ohms and resistors R2 were 295 ohms. Inasmuch as this is a network tuned to a certain frequency, our resistors can't have any deviation from the design value. For this reason, variable resistors were used. The procedure is to set the two resistors R2 to the correct value with an ohmmeter and then to adjust R1 while the network is in the circuit.

168

Figure 9-5 shows the arrangement for making measurements in the presence of the power line voltage. The Twin-T network of Fig. 9-4 is connected between the power line and the input of an oscilloscope. Then R1 is adjusted to minimize the deflection caused by the 60 Hz voltage. When the deflection is minimized, the two resistors R2 can be altered slightly to further minimize the deflection. With a little care, the 60 Hz voltage can be reduced to a very low level.

The filter acts by balancing out the voltage that reaches the output through the two branches of the circuit. It is very important therefore that there be little or no stray coupling between the input and output sides of the network. At 60 Hz this isn't much of a problem and adequate isolation can usually be obtained by just keeping the input and output leads well separated.

When the circuit of Fig. 9-5 is first used, the display is very interesting. All sorts of things that are not normally visible on an oscilloscope begin to show up on the screen. There is usually a lot of noise on the line and, as we noted earlier, there is often one or more harmonics of the power line frequency.

POWER LINE MEASUREMENTS

There are two types of measurements on power lines that help in getting a carrier-current system working properly. The first

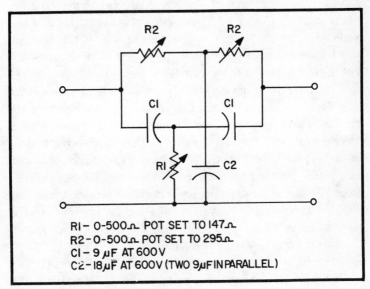

RI – 0-500 Ω POT SET TO 147 Ω
R2 – 0-500 Ω POT SET TO 295 Ω
CI – 9 μF AT 600 V
C2 – 18 μF AT 600 V (TWO 9 μF IN PARALLEL)

Fig. 9-4. Practical 60 Hz Twin-T network.

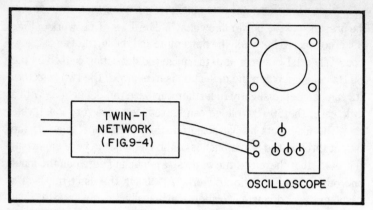

Fig. 9-5. Signal tracing set up.

consists in simply looking at the power line to see what is present other than the normal 60 Hz voltage. There will usually be some noise, but if it isn't too great it won't cause problems. If the noise is excessive it can interfere with the receiver or receivers in the system. Steps will have to be taken to find what is causing the interference and to eliminate it.

Sometimes most unusual and suprising things will be found on an ordinary power line. In one case, an 80 kHz signal of nearly 2 volts was found on a power line. The signal was intermittent and the source was never located. A carrier-current system at 100 kHz was made to operate satisfactorily, however, in spite of this spurious signal. In searching the power line, the goal is to locate any signal that might interfere with the carrier current system.

The next type of measurement is ordinary signal tracing of the type done in radio servicing. The carrier-current transmitter is connected to the power line and the signal is traced throughout the building with the test setup of Fig. 9-5. The carrier-current signal will be very weak, but it should have a level of a few millivolts at least and should produce a measurable deflection on a sensitive oscilloscope. By careful signal tracing, not only can it be ascertained that there is enough signal above the noise at the place where the receiver or receivers will be installed, but any appliance that is shunting out the signal can be located.

The signal tracing procedure is to turn off or disconnect as many of the things that are normally connected across the power line as possible. Devices such as furnaces which are permanently connected to the line can be disconnected by removing fuses or opening circuit breakers. When this is done there will usually be plenty of signal all through the building except possibly between things connected to different sides of the power line. This situation can usually be remedied by connecting a capacitor from one side of the line to the other as shown in Fig. 9-6. The 60 Hz leakage through this capacitor is only about 4 mA.

After the signal has been found to be adequate with no load on the wiring, the various lights and appliances are turned on one at a time to see what they do to the signal. They will almost always cause some attenuation, but this is to be expected. Problems arise when connecting a particular appliance makes the carrier-current signal disappear completely. When this happens, something must be done or the system will never work properly.

POWER LINE CONDITIONING

In a large commercial installation such as an electric utility, power lines that are to carry signals in addition to power are conditioned to prevent the signal from being shunted through devices connected to the line. This cannot be carried out on the same scale in a small installation simply because of the expense involved. If much line conditioning is required, it is usually better to use some other sort of control system.

There are a few things that can be done in a small building, particularly when all of the trouble is caused by one or two devices that are connected across the line. Before we discuss the steps that can be taken, let's look at just how an appliance connected to a power line can shunt out a signal of say 100 kHz.

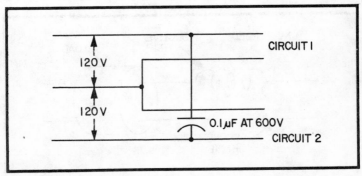

Fig. 9-6. Connecting a capacitor from side of a power line to the other.

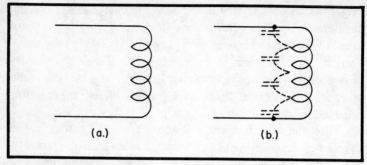

Fig. 9-7. Effect of distributed capacitance in an inductance.

Resistance loads such as lamps will simply provide a low impedance path for rf signals as they do for the regular 60 Hz signals. Fortunately, the impedance of a lamp at rf is usually quite a bit higher than at 60 Hz. This reduces the effect to some extent. Motors and transformers are by nature inductive devices and as such will have a higher impedance at rf. The problem in these devices is usually the distributed capacitance of the windings.

Figure 9-7 shows the situation. Although the element in the figure is an inductance, there is a great deal of capacitance between the individual turns of the winding. Thus, instead of looking like an inductance, the device looks electrically like both an inductance and a capacitance. There are two approaches to minimizing this problem—using a lower carrier frequency, and installing a series inductance in the line.

The simplest approach to the problem is to temporarily make the frequency of the carrier current transmitter variable. It is then tuned for maximum signal at the receiver location. Sometimes this

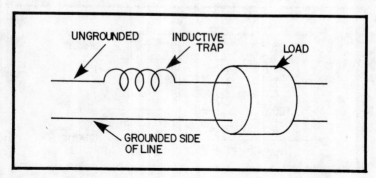

Fig. 9-8. Installing an inductive trap.

will produce enough of a change in level to make an otherwise unworkable system practical.

The other approach to the problem is to install inductance in series with the ungrounded side of the power line. This is shown in Fig. 9-8. If the device that is causing the problem doesn't draw much current, this approach is practical. If the device draws a large current, however, building the series inductance can be a problem because the wire must be large enough to safely carry the full current of the appliance. In Fig. 9-9 an inductance is made by winding as many turns as practical of the same gauge of wire that is used in the appliance around the ferrite core taken from a radio loopstick antenna. Again, in low current appliances this approach can be very effective. In high current appliances the wire is so large that not many turns can be placed on the coil with the result that its inductance will be lower and it will be less effective.

This whole business of conditioning a power line to carry rf signals tends to be more of an art than a science and it is a challenge to the experimenter.

The worst enemy of the carrier-current system is the radio interference filter that may take one of the forms shown in Fig. 9-10. This filter is installed in the inputs to an appliance to keep it from causing interference or from being susceptible to interference. In almost all cases there is a capacitor across the power line. The basic function of this capacitor is to provide a short circuit for any rf that might be on the power line. Of course, it will also look like a short circuit to the carrier current signal.

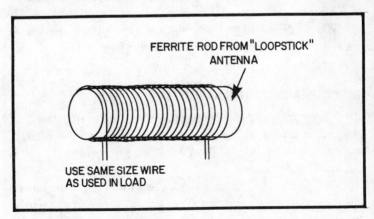

Fig. 9-9. Homemade inductive trap.

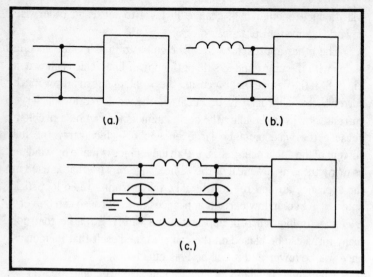

Fig. 9-10. Typical radio interference filters.

There are several steps that can be taken. Sometimes the filter is merely a precautionary measure and removing it will not cause serious interference problems. Usually, the solution is not this simple.

The filters used in commerically available appliances are usually available. When this is true, the same type of inductor used in the filter can be added in series with the line side of the filter as shown in Fig. 9-11. This will often minimize the influence of the filter on the rf signal on the line without loosing the interference reducing features of the filter.

When a filter of the type shown in Fig. 9-10c is used, it is sometimes satisfactory to simply remove the filters that are closest to the power line.

THE CARRIER CURRENT TRANSMITTER

There are many different approaches that can be taken to designing a carrier current transmitter. There is no fixed design approach because no two power lines will have exactly the same characteristics at rf.

One problem in designing the transmitter is to get the carrier current signal on the power line. The fact is that not only do we have no idea of what the rf impedances of the power line is, but the

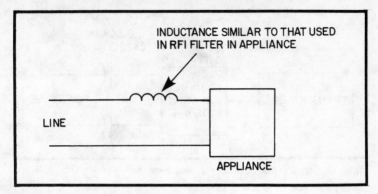

Fig. 9-11. Using an inductor from an RFI filter.

impedance will change when various things that are connected to the power line are switched on and off.

The conventional approach to connecting a transmitter to a load is to do some sort of impedance matching. The rule that governs this is that maximum power transfer occurs when the impedance of the load is matched to the impedance of the transmitter.

This brings up an aspect of power transfer that isn't widely appreciated. Figure 9-12a shows the equivalent circuit of a transmitter connected to a load. We know that we will get maximum power into the load when the impedance of the load is matched to that of the transmitter. Figure 9-12b shows the situation corresponding to connecting a carrier current transmitter to a power line. Here we have no control of the impedance of the load. In fact, we don't even

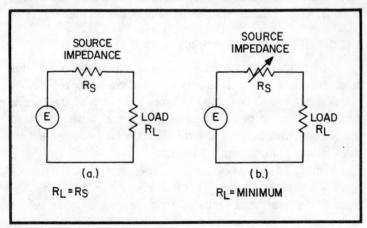

Fig. 9-12. Obtaining maximum power transfer.

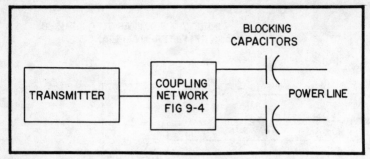

Fig. 9-13. Coupling the transmitter to the power line.

know what it is. In the design stage, at least, we do have control over the output impedance. In the circuit of Fig. 9-12b maximum power transfer *does not occur when the transmitter and the load are matched. Maximum power is delivered to the load when the output impedance of the transmitter is made as low as possible*. This means simply that in any given installation, regardless of what the impedance of the power line happens to be, the system will work better as the impedance of the transmitter output is made lower.

Another important consideration in connecting a transmitter to a power line is the fact that in addition to the rf getting onto the power line, the 60 Hz power will try to get into the transmitter. One approach to minimizing this is to use the Twin T network described in an earlier paragraph to reject the 60 Hz power while allowing the rf signal to pass. Figure 9-13 shows such a coupling network. As in the test setup of Fig. 9-5 the Twin T is adjusted to minimize the amount of 60 Hz voltage on the transmitter side of the network.

THE CARRIER FREQUENCY SOURCE

Generating a low frequency signal with a conventional oscillator is inconvenient because of the large values of inductance required. A practical alternative is to use a Type 555 timer as the carrier generator. Although this circuit generates a square wave rather than a sine wave, this isn't particularly objectionable because the harmonics can be filtered out rather easily.

Figure 9-14 shows a circuit for a carrier generator. The components are arranged so that the output will be a square wave. The frequency is determined by the values of capacitor C1 and resistors R1 and R2. The values of these components can be determined from the equation:

$$f = \frac{1.44}{(R1 + 2R2)\ C}$$

Using the values given in Fig. 9-14 the frequency can be adjusted from over 100 kHz to as low as 1 kHz.

As shown in the Fig. 9-14, pin 5 of the timer can be used to frequency modulate the carrier. Actually, the modulation is more like frequency shift keying than conventional FM and it seems to perform better than conventional AM in a system where the signal is propagated along the power line. Furthermore it eliminates problems that might otherwise arise in connection with modulating the low frequency carrier.

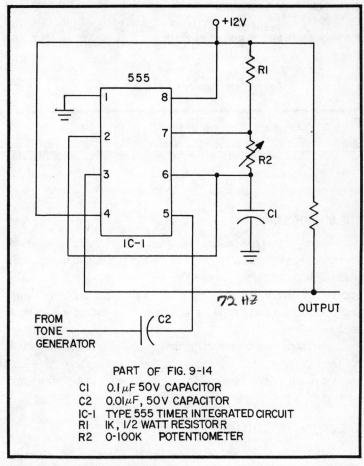

PART OF FIG. 9-14

C1   0.1 μF 50V CAPACITOR
C2   0.01μF, 50V CAPACITOR
IC-1 TYPE 555 TIMER INTEGRATED CIRCUIT
R1   1K, 1/2 WATT RESISTOR R
R2   0-100K   POTENTIOMETER

Fig. 9-14. Carrier frequency generator.

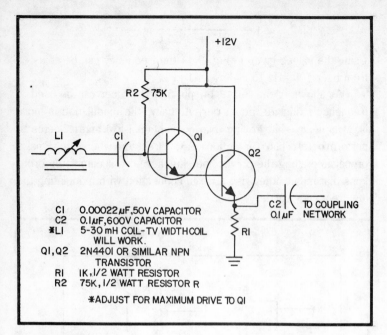

Fig. 9-15. Output stage.

The fact that the carrier frequency is very low means that the tone frequency used for modulation can't be too high. Again, the frequencies used in the dual tone dialing system are ideal.

THE OUTPUT STAGE

As we pointed out earlier, one of the important requirements of the output stage is that it have a very low internal output impedance. The Darlington stage shown in Fig. 9-15 will provide this characteristic. Of course, it is possible to use a similar arrangement with amplitude modulation, but the FM arrangement usually works very well.

The transistors shown in the figure will provide enough output for small buildings, but it is probable that some applications might require much more power. Larger transistors can be used with corresponding changes in components.

THE COMPLETE TRANSMITTER

Figure 9-16 shows a complete carrier current transmitter. The stages are those that we discussed earlier. The Type 555 timer

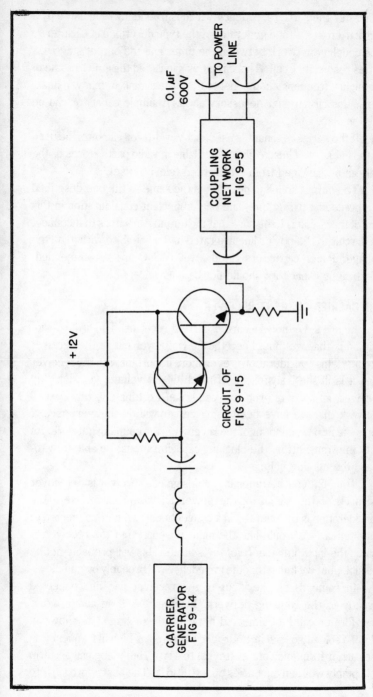

Fig. 9-16. Complete carrier current transmitter.

179

generates the carrier frequency which is frequency modulated by a signal from a two-tone generator of the type described in Chapter 7. The coupling network between the timer and the output stage is a series resonant circuit. This circuit was included to remove some of the higher frequency components from the signal. In many installations, this circuit isn't necessary at all. A simple capacitor can be used in place of it.

If the series resonant is used, a TV width coil can often be used as the inductor. These coils are available at radio parts stores or the coil can be salvaged from a discarded television set.

The output coupling network is the same as the one described in a preceding paragraph. The only important consideration in this part of the circuit is to be sure that the output capacitors that connect the circuit to the power line are rated high enough to withstand the voltage. Paper capacitors rated at 600V DC are recommended. Electrolytic capacitors should not be used.

THE CARRIER CURRENT RECEIVER

Figure 9-17 shows a carrier current receiver. The first consideration is that we must keep the 60 Hz power out of the receiver circuits. This consideration is even more important with the receiver than it is in the transmitter. If much 60 Hz voltage gets into the receiver, all sorts of problems can arise. Fortunately, the Twin-T network that we have been using earlier will do this very nicely. Thus the first component of the receiver is the coupling network. As with the transmitter, the blocking capacitors must be able to withstand the full line voltage.

The next consideration in the design of a receiver is to recover as much of the signal on the power line as possible. Here the consideration is the same as in a conventional receiver connected to an antenna. We would like the input impedance of the receiver to match the input impedance of the source. This will probably not be possible, but we know that the impedance will probably be quite low.

In some cases the input of the receiver can be connected directly to the coupling network. In other cases an audio output transformer can be connected between these stages as shown in Fig. 9-17b. Audio transformers are not rated as high in frequency as the carrier frequency of a system of this type, but many will perform acceptably well at frequencies up to 100 kHz. The best approach is to

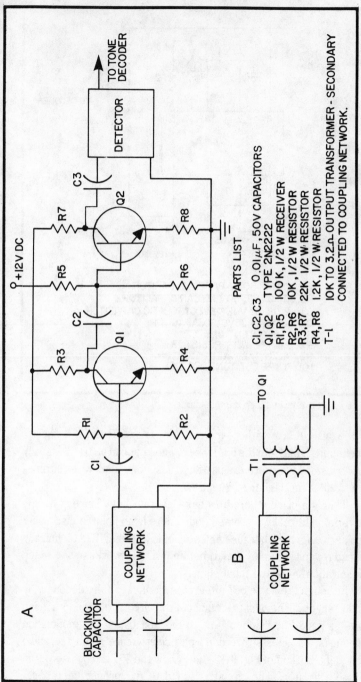

Fig. 9-17. Carrier current receiver.

PARTS LIST

C1, C2, C3   0.01 μF, 50V CAPACITORS
Q1, Q2   TYPE 2N2222
R1, R5   100K, 1/2 W RECEIVER
R2, R6   10K, 1/2 W RESISTOR
R3, R7   22K, 1/2 W RESISTOR
R4, R8   1.2K, 1/2 W RESISTOR
T-1   10K TO 3.2 Ω OUTPUT TRANSFORMER - SECONDARY
              CONNECTED TO COUPLING NETWORK.

181

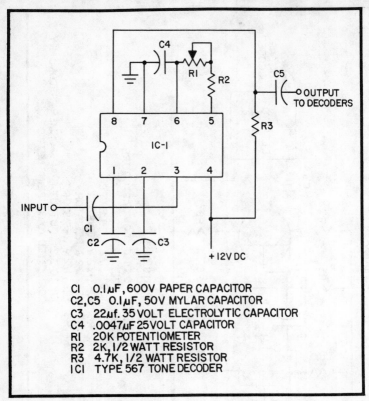

Fig. 9-18. Phase-locked loop demodulator.

C1   0.1μF, 600V PAPER CAPACITOR
C2,C5   0.1μF, 50V MYLAR CAPACITOR
C3   22μf, 35 VOLT ELECTROLYTIC CAPACITOR
C4   .0047μF 25 VOLT CAPACITOR
R1   20K POTENTIOMETER
R2   2K, 1/2 WATT RESISTOR
R3   4.7K, 1/2 WATT RESISTOR
IC1   TYPE 567 TONE DECODER

try the system first without the matching transformer. If it doesn't get enough signal from the line, the transformer can be tried. Usually the very small transformers used in transistor radios work surprisingly well at the higher frequencies.

This is a place where a rather elaborate impedance matching network might really do some good. There is no point in discussing such networks here because no two would be alike. The technically inclined reader will enjoy experimenting with networks to see which configuration gives the most signal.

Although the Type 567 integrated circuit is a tone decoder, by making slight changes in the values of the circuit components it can serve as good FM demodulator. In this application the modulation will consist of tones so we don't particularly care about the audio quality. The circuit of Fig. 9-18 shows how the 567 can be used as an FM demodulator. The circuit is tuned to the proper center frequency by means of the pot R1. With the values shown the circuit should tune from about 10 kHz to over 100 kHz. The output circuit may be connected to other 567s connected as tone decoders as described in Chapter 7.

In the parts list, capacitor C1 is specified as having a voltage rating of 600V so that the circuit can be connected to the power line. If isolation is provided ahead of the demodulator in a coupling network, a capacitor having a lower voltage rating can be used. The simplest carrier-current system is one that uses only a carrier, without modulations, to turn something on or off. Such a system is shown in Fig. 9-19. Here the carrier is turned on to send a control signal. No detector circuit is required at the receiver, because a Type 567 tone decoder can be used after the amplifier to decode the carrier itself. The RC frequency determining elements must be chosen so that the circuit will respond to the carrier frequency. See Chapter 7 for design details. A toggle arrangement can be used so that the first transmission of the carrier will turn the controlled device on and the following transmission will turn it off.

This system is very useful as an addition to some other sort of control system where most of the devices are connected to the system through control wiring, but one device is located where it is very difficult to run control wires. Another good application for a system of this type is for a portable application. There are applications where it is desired to control something that may be moved around somewhat frequently. For example, we might have a portable electric fan that may be used in various parts of the house at different times. The fan, being AC operated, will always be connected to the power line. If a carrier-current system is used for control, there will be no need to run temporary control wiring to each location where it might be used.

In some applications there is no substitute for a complete carrier current system. Some buildings, particularly those of

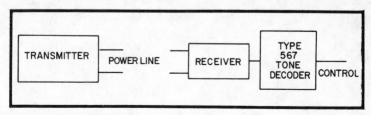

Fig. 9-19. Control system using only carrier for control.

masonry construction, simply have no place for running control wiring. If a control system is to be used at all, the power lines must be used as the control wiring. In such an application, a carrier current system is justified, and it is usually worthwhile to do the necessary debugging to get it working properly.

As we pointed out earlier, it is rare that a system of this type can be built and simply installed and expected to work properly. The fact is that there is a completely unknown situation—the power line wiring—right in the middle of the system. It is usually necessary to do quite a bit of signal tracing and debugging of the system to get the system operating properly.

10
Time Controlled Systems

It has been said that the American people live by the clock. A great deal of human activity is indeed predicated on the time of day. Most people start their day on a signal from an alarm clock and spend the rest of the day following some sort of schedule that is closely geared to time.

Many of the operations that are performed by human beings at definite times can be automated so that the operations will be initiated by an electronic signal at the proper time. Of course, this is common in industrial plants, but it is rarely applied in the home or small business.

The most obvious function that can be automated with an automatic timing system is the alarm clock itself. A master clock that is used to control other functions can easily be connected to sound an alarm that will wake a person in the morning. In addition to the waking function the system can also control a coffee maker and a toaster. The timing can be set so that the toast and coffee will be ready when the alarm sounds. Although this application is trivial, it certainly makes getting up in the morning much more pleasant.

A more useful time controlled device is a timed thermostat. This is like a conventional thermostat that controls a heating or air conditioning system, except that it provides different temperature settings for different times of day. For example the thermostat could be set to keep a home at a normal comfortable temperature during

Fig. 10-1. Radio Shack Type 63-864 timer.

daytime hours when the family is awake and moving about. It could also be arranged to automatically reduce the temperature at some convenient time such as 11:00 pm when the family normally retires. The second setting of the thermostat would be at a lower temperature which would allow the house to cool while nobody was awake. An early morning signal, before the alarm clock signal, could be used to switch the thermostat back to the normal temperature so that the house would be warm by the time the family woke up. With a little imagination many different areas of application can be found for a time controlled system. A typical application might be to use the system to control the sound level from the telephone bell or doorbell. The system could be set to provide a comfortable but noticeable volume during the day when everyone was awake. The timer signal could increase the sound level from the bell at night when everyone was asleep so that the bell would be loud enough to awaken someone.

ELECTROMECHANICAL TIMERS

In this day of marvelous electronic devices, it may seem enigmatic to describe the use of electromechanical timers in a modern control system. The reason is simply that electromechanical timers are very reliable and of course they are extremely easy to install. Using a complex electronic circuit to do what can be done with a simple electromechanical unit is like driving the proverbial nail with a sledge hammer.

186

Figure 10-1 shows a simple electric timer of the type available at hardware and discount stores. It is simple, and will serve any function where a single device is to be turned on and off once a day. If several devices need to be turned on and off at various times during the day, the electromectanical timer shouldn't be dismissed without further investigation. There are many electromechanical timers used in industrial applications that will handle a large number of devices. These timers, shown in Fig. 10-2, consist of a motor which drives an assembly of cams which, in turn, operate snap actions. Often a device like this can be picked up at a surplus outlet. If the original time cycle is something other than a 24-hour period, the device can sometimes be changed for a 24-hour cycle by simply replacing the synchronous motor with one that makes one revolution in 24 hours. Motors of this type are often available as surplus items.

THE TIME BASE

Accuracy is extremely important in a timing system that runs continuously. The reason is that any error tends to accumulate as time passes. Inasmuch as a system will probably remain in operation for many years, even the smallest error can become significant if it is always in the same direction. The situation is improved somewhat by the fact that random errors are as likely to be in one direction as the other and tend to cancel out to some extent.

The significance of the accuracy of the time base in a timing system can be seen from the following example. Suppose that the time base of a system had a consistent error of one second per day. At the end of 2 months the whole system would be off by a minute. This isn't very important, but at the end of the year it would be off by 6 minutes and the error would accumulate at the rate of 6 minutes per year. For this reason, the time base of a system must be very accurate, or provision must be made for checking and resetting the system at periodic intervals.

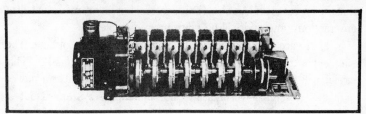

Fig. 10-2. Industrial timer with 8 circuits.

Interestingly enough, one of the most accurate sources of timing signals is the ordinary 60 Hz power line. The reason is not that the frequency of the power line is held to such close tolerances; it is that the utility companies regularly correct errors that may have accumulated because of changes in the power line frequency during the day.

Whenever the load of a power system increases, the generators tend to slow down until the system can readjust itself to the new load requirement. Similarly, when the load decreases, the generators will tend to speed up. These changes, although very small, will change the power line frequency enough to make an electric clock run fast or slow, depending on the direction of the frequency shift. To keep the electric clocks accurate, the utility company periodically adjusts the frequency in such a direction as to correct the error. This change is usually made during the early hours of the morning so that by the time the working day starts the clocks are all at the correct time. This is why it isn't necessary to reset an electric clock unless there has been a power outage.

Thus the 60 Hz power line is probably the best source of timing information for a time controlled system. The only disadvantage is that the system gets completely out of synchronization whenever there is a power outage. This may or may not be significant. The fact that the timing source fails during a power outage is not in itself significant in a system that depends on the power line for its operating power. Nothing else in the system will work without power, so it isn't important that the time base won't work either. The problem arises from the fact that the timing of the system must be reset when the power is restored. This problem can be conpounded by the fact that there may be no one around to reset the system and it will start to run when power is restored. If significant functions are controlled by the system everything will happen at the wrong time. If an electromechanical timer is used, it will be slow by the length of time that the power is off. An electronic system, however, may start up at any random time and will be quite unpredictable.

There are two approaches to handling the power outage. The simplest is the power failure indicator shown in Fig. 10-3. This circuit will flash a light when the power is restored after a power outage. The signal will indicate that the timing system is out of synchronization.

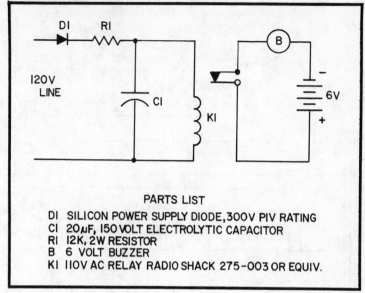

PARTS LIST
DI SILICON POWER SUPPLY DIODE, 300V PIV RATING
CI 20µF, 150 VOLT ELECTROLYTIC CAPACITOR
RI 12K, 2W RESISTOR
B   6 VOLT BUZZER
KI IIOV AC RELAY RADIO SHACK 275-003 OR EQUIV.

Fig. 10-3. Power failure alarm.

The other approach is to include in the system a battery operated crystal oscillator that will provide timing signals when the 60 Hz
signal from the power line is not available. All the battery powered
part of the system has to do is keep time until power is restored. It
needn't be a very big device, because the rest of the control system
will probably not be working anyway.

ELECTRONIC TIME BASES

Most home control functions are not particularly critical. If the
timed event occurs within a minute or so of the selected time no
harm will be done. Thus if a time base of a system produces one
pulse per minute, the system timing will be adequate. We stated
earlier that the 60 Hz power line is an excellent source of timing
information. The problem, then, is to start with an input of 60 Hz and
produce an output of one pulse per minute. One approach is to simply
divide the power line frequency by 3600 with a string of binary
dividers. Fortunately, this isn't necessary because there is a single
integrated circuit that will do the job. It is the Motorola Type
MC-14040 Counter.

Figure 10-4a shows a block diagram and pin connections of the
Type 14040 counter. As shown in the figure, this IC consists of

189

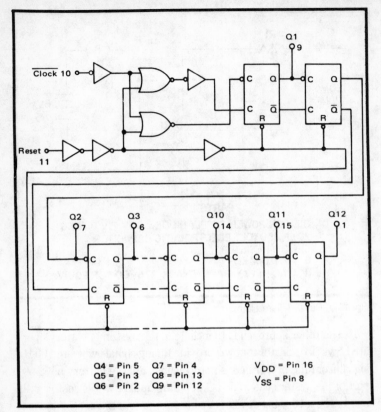

Fig. 10-4. Block diagram and pin connections of Motorola Type MC 14040 counter.

waveshaping circuits and a string of 12 divider stages. The outputs of the dividers are brought out to pins labeled Q1 through Q12. There are two separate input pins. Pin 10 is where the signal to be counted is fed into the circuit. This signal is usually called a clock signal or simply a clock. Pin 11 is a reset pin that will reset all of the counter outputs to zero when it is brought to a high logic level.

The Type MC-14040 uses CMOS technology so it draws very little power and is reasonably immune to noise. The supply voltage can be anywhere between 3 and 16 volts. If it is desired to keep the internal power dissipation low, as it would be in a battery powered stage, the supply voltage can be held to about 5 volts.

The operation of the stage can be understood by following the timing diagram shown in Fig. 10-5. The top line in this diagram is the timing pulse that is to be counted or divided to a lower pulse

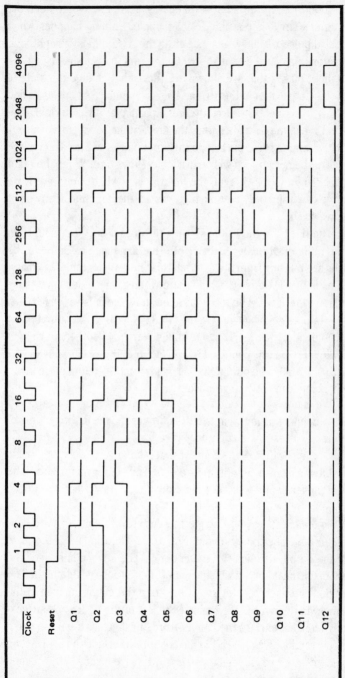

Fig. 10-5. Timing diagram of Type MC 14040 counter.

repetition rate. Note that several pulses are not shown to keep the diagram as small as possible. For example, nothing happens when the third pulse reaches the input after the reset pin is brought low. Therefore this pulse is not shown on the timing diagram. In fact the only input pulses that are shown are those that change the state of the following divider in the string. This is enough to enable us to find out how to connect the circuit to get a particular division ratio. The second line in the diagram shows the state of the reset pin. All it says is that when the reset pin is high the input pulses will not be counted and all of the output pins will be in a logical low state. When the reset pin is brought to a low level, the input pulses will be counted. Note that we can bring the reset pin high to stop the counting at any point that we wish.

The third line shows the state of the output pin Q1 as a function of the number of input pulses. Note that inasmuch as some of the pulses are missing from the diagram this line is not complete. What it tells us is that Q1 goes to a high state when every other input pulse drops to a low level. The following lines show the pulse at which each of the outputs goes high for the first time. Note that the output pin of the last divider stage, Q12, goes high when the 2048th input goes to a low level. Don't be confused by the fact that the twelfth power of 2 is actually 4096. The output pins all go low on the 4096th pulse as we might expect in a 12 stage divider.

We can use the counter to divide by any integer up to 4096 by merely connecting logic gates to its outputs and using the outputs of the gates to reset the counter.

One application of the counter that is useful is to develop a pulse train having a pulse repetition rate of one pulse per minute from the 60 Hz power line. To this we must divide the power line frequency by 3600.

The easiest way to decode the output of a counter is to construct a table like that shown in Fig. 10-6. The top line of the table lists the output terminals of the counter. The second line gives the number of the count at which each of the outputs will go high. The way that we use these figures to decode the count is given in the example in the figure. We start at the left of the figure and successively subtract each of the counts from our divisor. If we can make the subtraction, we place a 1 in the column. If making the subtraction would give us a result less than zero, we place a zero in the column and go to the next count.

OUTPUT TERMINAL	Q12	Q11	Q10	Q9	Q8	Q7	Q6	Q5	Q4	Q3	Q2	Q1
COUNT	2048	1024	512	256	128	64	32	16	8	4	2	1
	1	1	1	0	0	0	0	1	0	0	0	0

EXAMPLE

```
3600
2048
────
1552
1024
────
 528
 512
────
  16
  16
────
   0
```

Fig. 10-6. Chart for calculating binary division.

In the example, we want to divide by 3600. Starting at the left column in our table, we find that the first number is 2048. We will therefore place a 1 in this column of the chart and subtract the number 2048 from our desired divisor of 3600. This gives us 1552. Now we will go to the second column of the chart and find that the number corresponding to this column is 1024. We will then place a one in this column and subtract its number 1024, from our remainder of 1552. This gives us a new remainder of 528. Going to the next column, we subtract 512 from our remainder of 528 and put a one in the column on the chart. This gives a new remainder of 16. The next column is 256 which is greater than our remainder of 16, so we place a zero in the column and go on. The next column is also too large so we place a zero on the chart. The following two columns corresponding to 64 and 32 are also too large so we again place zeroes in these columns on the chart. The next column is a lucky hit being exactly equal to our remainder of 16 so we place a one on the chart at this space. We have now found our desired number, so we place zeroes in the remaining columns of the chart.

Looking at the chart we have made up we see that 3600 input pulses have occurred when the outputs at Q5, Q10 and Q11 and Q12 are all high. What we have to do now is to detect this condition, and form an output pulse to reset the counter.

Figure 10-7 shows the diagram of a circuit that will do this for us. The inputs of a four-input NAND gate are connected to the pins that will go high when 3600 pulses have entered the counter. When this condition is reached and not before then, the output of the NAND gate will go low. We actually want a high output so another 4-input NAND gate is connected as an inverter to make the output

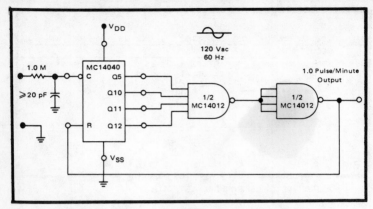

Fig. 10-7. Circuit that produces one pulse per minute from 60 Hz line.

positive. This positive output is fed back to the reset pin on the counter to reset it. Thus the circuit of Fig. 10-7 will provide one output pulse for every 3600 input pulses. If the input is a 60 Hz signal, the output will be a one pulse per minute signal. This is exactly what we want to drive a timing system.

The second 4-input NAND gate was used as an inverter in the circuit of Fig. 10-7 simply because there happened to be two gates of this type in the integrated circuit. Any other type of inverter could have been used.

COUNTING MINUTES

Now that we have a source of pulses that occur every minute, all we have to do to tell the time of day is to find some way to count the pulses from some arbitrary starting point. The system that uses the least hardware is one where the minutes are counted as binary numbers. Very little hardware is required, but it is difficult to read the time. If we count using the Binary Coded Decimal (BCD) system, more hardware will be required, but the numbers will be easier to interpret. The reader will want to make his own choice, so we will describe both systems.

Figure 10-8 shows a binary time keeping system. This circuit is very similar to that of Fig. 10-7 that we used to get our one minute pulses from the 60 Hz power line frequency. In this case we count up to 1440 and then reset the counter.

The reason is that there are 60 minutes in each hour and 24 hours in each day giving us 1440 minutes in a day. We want the

194

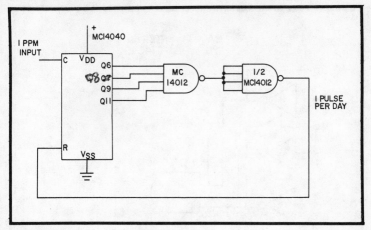

Fig. 10-8. Circuit to count minutes.

counter to count minutes, starting preferably at midnight, and reset again at midnight the next day. To determine which outputs of the counter to use to get a count of 1440, we use the same procedure that we used to set the one pulse per minute counter. This gives the result that we must detect the condition where outputs Q6, Q8, Q9, and Q11 are high. The fact that this will give us a count of 1440 is shown in Fig. 10-9. Although it isn't shown in Fig. 10-8, all of the outputs, Q1 through Q11, are brought out to leads. We will use the signals on these leads to tell the time of day. Of course, we must do this by expressing the time as a true binary number representing the number of minutes that have passed since midnight. This is a little troublesome, but inasmuch as most time controlled systems are not reset very often, it may be worth the trouble to save hardware.

DECODING THE TIME

In order to make something turn on or off at a predetermined time we must be able to decode the time on our 11-bit bus consisting of lines from pins Q1 through Q11. This is really quite simple. A setup

Fig. 10-9. Counting to 1440.

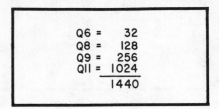

Q6	=	32
Q8	=	128
Q9	=	256
Q11	=	1024
		1440

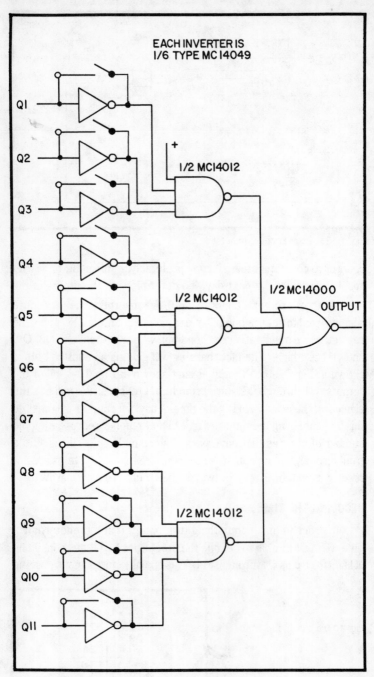

Fig. 10-10. Circuit that will decode any minute in the day.

that will decode any time of day is shown in Fig. 10-10. The part of the circuit consisting of three 4-input NAND gates and one 3-input NOR gate is actually a 12-input AND gate. The output will go high when and only when all of the inputs are high. The trick is to make the inputs all high at the desired time. This is done by switching inverters into those leads that will be low at the desired time.

For example, suppose that we want to turn something on at 6:30 in the morning. This is six and a half hours after midnight, or in terms of minutes, 390 minutes after midnight. The binary number corresponding to 390 is shown in Fig. 10-11. The number is 00110000110 and is found when the various outputs have the state shown in Fig. 10-11. All of the outputs that are zero at this time are connected to the gate through inverters so that the inputs to the 12-input AND gate will all be high exactly 390 minutes after midnight. A similar arrangement can be made for any other time of day.

There are many ways that the decoding logic can be simplified. Much depends on how fine the increments of time must be in a particular application. Suppose, for example, that we were willing to divide the day into 8-minute intervals instead of into hours and minutes. We could then disregard the state of the first three least significant outputs of the counter. This would mean that we only had to decode an 8-bit binary number.

Decoding an 8-bit number is easy because a single integrated circuit gate can be used for the purpose. The Motorola Type 14501 Tripple gate, shown in Fig. 10-12, can be made into an 8-input gate by connecting pin 10 to pin 12 and pin 11 to pin 13 as shown by the dotted lines in the figure. The gate includes an inverter so if the output is taken from pin 14 it will behave as an AND gate, and if the output is taken from pin 15 it will behave as a NAND gate.

6:30 = (6 X 60) + 30 = 390 MINUTES

$$Q9 = 256$$
$$Q8 = 128$$
$$Q3 = 4$$
$$Q2 = 2$$
$$\overline{390}$$

Fig. 10-11. Decoding 6:30 AM (390 minutes after 12 PM).

197

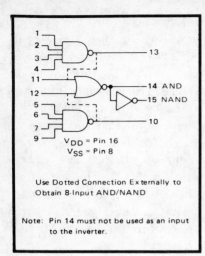

Fig. 10-12. Motorola Type 14501 8-input gate.

V_DD = Pin 16
V_SS = Pin 8

Use Dotted Connection Externally to Obtain 8-Input AND/NAND

Note: Pin 14 must not be used as an input to the inverter.

Of course, it will still be necessary to use inverters ahead of the inputs in the lines that will be low at the desired time of day.

Another approach to simplifying the timing system is to use much coarser increments of time than the minute. There are few things in the home that have to be controlled to the minute. Often a pulse once every 15 minutes will be all that is required to provide the desired time control. Figure 10-13 shows another counter similar to the ones that we have discussed that counts minutes and provides an output every 15 minutes. The outputs of this counter will be in 15-minute intervals and must be decoded accordingly. There are 24

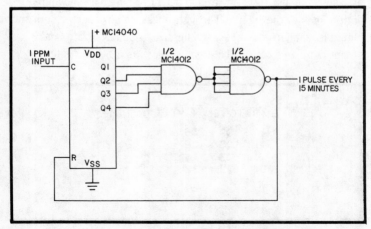

Fig. 10-13. Circuit to produce one pulse every 15 minutes.

hours in a day, and there are 4 15-minute intervals in each hour so there are $24 \times 4 = 96$ 15-minute intervals in a day. Thus the counter shown in Fig. 10-13 will deliver 96 output pulses in a day.

READING THE TIME

The time systems described so far are reasonably simple and arriving at the proper decoding isn't very difficult once one has gained a little experience with it. There is one big limitation to the system so far and that is there is no way of telling what time the system says it is at any given moment. The system uses the binary number system, so we must be able to read the time as a binary number. There are two approaches to the problem. One is to use a series of circuits like that of Fig. 10-14 that will drive a light emitting diode (LED) from each line of the binary number. We then merely have to add up the number corresponding to each line to tell what time the system says it is. Another approach is to use a voltmeter to see which lines are high and which are low.

In either case we need a way to set the time of day so that the system indicates the correct time. Of course one way is to wait until midnight and reset the whole thing. This will work, but it has its limitations. Another approach is to drive the counter that is doing the

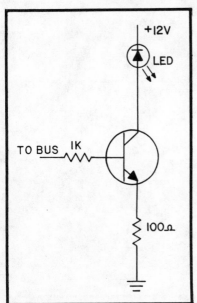

Fig. 10-14. Transistor LED driver.

decoding at some faster rate. This can be done by temporarily jumping one of the other pins of the 60 Hz counter to the output. Here the output from the counter is one pulse per minute, but the outputs of the other pins have much faster pulse rates. In fact pin Q1 of the counter will have 60 pulses per second. This permits advancing the counting system by an hour in only a second. Moving on to the pin corresponding to output Q7 will give a slower rate of just over one pulse per second. Thus the rapid pulse rate can be used for coarse adjustment and the slower rate for fine adjustment.

11

Practical Home Control System

Some suggestions for things that can be controlled by an electronic system were given in Chapter 1. Of course, the number of things that can be controlled by such a system is not limited to these more obvious things. Almost anything can be controlled by an electronic system.

The first step in getting to a practical control system is to decide what is to be controlled. This is an important decision because history shows that the chances of a system ever being completed in a finite period of time varies inversely with the number and complexity of the things that are to be controlled. A simple system that will control a couple of lights can be designed, built, and installed over a weekend. A complex system that handles everything in the house will not be finished for months, if ever. An even longer period will elapse before it is working properly.

A good motto to use in designing a control system is expressed in the KISS formula. This means, "Keep It Simple, Stupid." The simpler the approach the more reliable the system will be. For this reason it is advisable to design a system that can later be expanded after the first few functions are working properly. This means putting in enough control switches and indicators so that functions can be added with a minimum of difficulty.

The second step, which should be obvious, is that no remote or automatic control should be applied to anything that doesn't work easily and reliably before the system is installed. Table 11-1 gives a list of things to check before attempting any control functions. There

Table 11-1. Things to Check before Installing Any Kind of Control System.

ELECTRIC DEVICES

1. ARE THE CONNECTIONS GOOD, AND WIRING ADEQUATE

2. IS THE SERVICE PROPERLY PROTECTED BY FUSES OR CIRCUIT
   BREAKERS

3. DOES THE DEVICE CAUSE RADIO OR TV INTERFERENCE. IF
   SO, IT MIGHT CAUSE INTERFERENCE TO CONTROL SYSTEM.

4. IS THE DEVICE SUSCEPTIBLE TO INTERFERENCE AND NOISE

5. HOW MUCH CURRENT DOES THE DEVICE DRAW   WILL THE
   PROPOSED CONTROL UNIT CARRY THIS CURRENT SAFELY

MECHANICAL DEVICES

1. DOES THE DEVICE FUNCTION PROPERLY

2. DO ALL MOVING PARTS MOVE SMOOTHLY WITHOUT
   BINDING

3. DOES FORCE HAVE TO BE APPLIED AT A PARTICULAR POINT
   FOR SMOOTH OPERATION

4. IF LUBRICATION IS REQUIRED, HAS THE DEVICE BEEN
   PROPERLY LUBRICATED

is nothing more frustrating than trying to open drapes or a window with a control system when the task is nearly impossible to perform manually.

This chapter includes descriptions of practical control systems and components that have actually been built and used in homes. Most of the things described here have been used in one particular home first and then later copied by others.

We will first describe the overall system, the things that it does, and the control switching arrangement. Then we will discuss the individual circuits and how they work.

THE MAIN CONTROL BOX

Figure 11-1 shows a DC control box that can provide control signals to operate 13 different devices. All of the switching arrange-

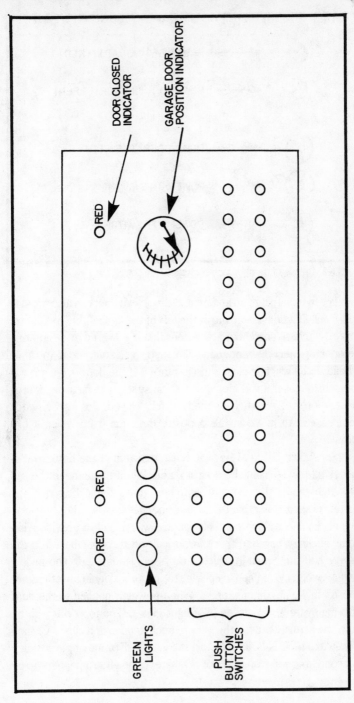

Fig. 11-1. Control box of typical home control system.

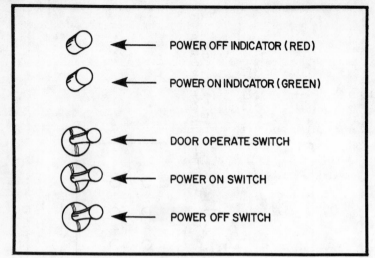

Fig. 11-2. Details of garage door control.

ments can turn things on and off and three switches can also provide additional functions. Starting at the left of the figure, the first row of three switches controls a garage door that is powered by an automatic door opener. This opener will cause the garage door to move whenever a switch is closed momentarily. If the door is open it will cause it to close; if it is closed it will cause it to open. If the door is somewhere between fully opened and fully closed, closing the switch will cause it to move in the opposite direction from which it last moved.

As shown in Fig. 11-2, the bottom switch in the first row removes all power from the door opener. When this is done the red pilot light above the row of switches glows, indicating that the system is secure and that the door cannot be opened either by a key, radio control, or a switch. This feature was added to provide additional security because there is no lock on the door between the garage and the house in which the system is installed. The middle switch when closed applies power to the door opening system, but doesn't make anything happen. When power is applied to the door opening system, the green pilot light glows. The top switch in the first row actually causes the garage door to open and close. Closing this switch momentarily will start the garage door moving. Closing it a second time will stop the door in whatever position it happens to be in.

The second row of switches, shown in Fig. 11-3, controls an intercom system. The bottom switch will remove all power from the intercom so that no one can eavesdrop. Pressing the middle switch momentarily applies power to the intercom. Under this condition, the outside speakers will act as microphones and the inside speakers will be listening. Thus any sound outside the doors will be heard inside the house. The top switch in this row is a push-to-talk switch. When it is closed, a spoken word inside the house will be heard outside near the doors.

The third row of switches controls the outside lights near the door. Closing the top switch momentarily causes all of the outside lights to be turned fully on. When this is done, a red pilot light will glow. Closing the middle switch momentarily dims the outside lights. This cause a green pilot light to glow. Finally closing the bottom switch momentarily turns the outside lights off completely and both pilot lights go out.

It was found desirable to be able to set the garage door at a known position between fully opened and fully closed for such purposes as to let pets go in and out of the garage without opening the door wide enough for a human being to enter. For this reason the indicating meter shown at the right side of the panel in Fig. 11-1 was added to the system. Details are shown in Fig. 11-5. The pointer of this meter has four positions corresponding to four different positions of the garage door. As an additional convenience the red pilot light of the meter was added. This pilot light glows when the door is

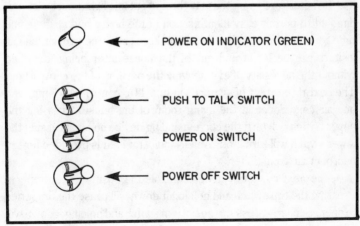

Fig. 11-3. Details of intercom control.

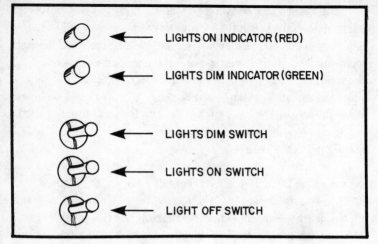

Fig. 11-4. Control of outside lights.

fully closed. Thus one can tell when the door is closed at a considerable distance from the control box.

The first three rows of switches in Fig. 11-1 together with the door position indicator, can be used to handle the admission of anyone to the house. When the doorbell rings the intercom is turned on and if it is after dark, the outside lights are also turned on. Communication is then established with the visitor using the push-to-talk switch. After the visitor has been satisfactorily identified he can be asked to enter through the garage door which can then be opened for him.

If visitors are expected at night, the outside lights can be left on, dimmed to permit easy identification of the house and at the same time conserve energy. By the use of three simple control circuits all visitors can be let in and out of the house after being identified without the necessity of ever leaving the position of the control box. The next three rows of switches in Fig. 11-3 control two lamps and one air conditioner in the living room of the house. Pressing the upper switch will turn the corresponding device on and pressing the lower switch will turn it off. An ON indicator light is provided for the bedroom air conditioner.

The next row of switches controls the position of a drapery. Pressing the top switch and holding it down will cause the draperies to open. Similarly pressing the bottom switch and holding it down will cause the draperies to close.

In this particular home installation the telephone was placed close to the control box so that it could be answered without leaving the central location of the system. Accordingly the next two switches on the control panel permit raising and lowering the volume of a stereo system. This not only permits adjusting the volume of the stereo to a convenient level but also makes it possible to mute the stereo when answering the telephone.

The five additional switches on the control box control other things in the house. Among the things that are controlled are a light, an air conditioner, and a fan in the bedroom of the house. In this particular installation the bedroom door could be seen at the end of a hall from the control point of the system. For this reason it wasn't felt that any indicators were necessary. It was found to be worthwhile, however, to add an indicator to give the state of the bedroom air conditioner.

A subsidiary control box for the system was installed in the bedroom. The panel and switching arrangement of this subsidiary control box are shown in Fig. 11-6. The first three rows of switches are the same as those on the main control box. This permits identifying and admitting visitors from the bedroom as well as from the main control point. The remaining rows of switches are on-off devices. They control the bedroom lights, air conditioner, fan, an FM radio for background music and one light in the living room.

It has been found that these two control boxes with a simple system using a DC control signal provide just about all the control that is needed for a handicapped person to live effectively in the home.

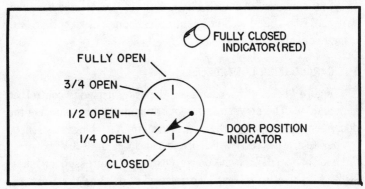

Fig. 11-5. Detail of garage door position indicator.

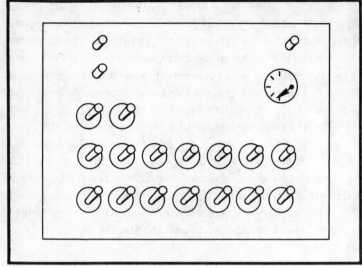

Fig. 11-6. Simplified bedroom control box.

THE POWER SUPPLIES

To keep things simple two separate DC power supplies were used in the system, one in connection with the main control box and a second one in connection with the auxiliary control box in a bedroom. The diagram of the power supply is given in Fig. 11-7. It is a simple power supply consisting of a transformer, filter capacitors and two rectifiers one to provide a positive signal and the other a negative signal. Filtering of the power supply isn't particularly critical and the drain on it is very small because the only current drawn except when things are being controlled is the current drawn by the indicators.

In the interest of avoiding large cluttered schematic diagrams, we will discuss the individual control functions one at a time with diagrams of their circuits.

THE GARAGE DOOR CONTROLLER

Figure 11-8 shows a diagram of the system used to control the garage door. The power to the door is controlled by plugging the door opening system into the control system. A latching relay is used to operate a larger relay which opens and closes the 120V power to the door opener. When the power ON switch is pressed a positive signal is applied to the latching relay. This latches the relay and energizes the larger relay which applies power to the system.

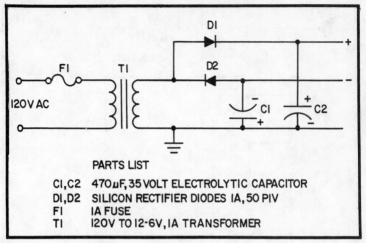

PARTS LIST

CI,C2   470μF, 35 VOLT ELECTROLYTIC CAPACITOR
DI,D2   SILICON RECTIFIER DIODES IA, 50 PIV
FI   IA FUSE
TI   120V TO 12-6V, IA TRANSFORMER

Fig. 11-7. Power supply for control system.

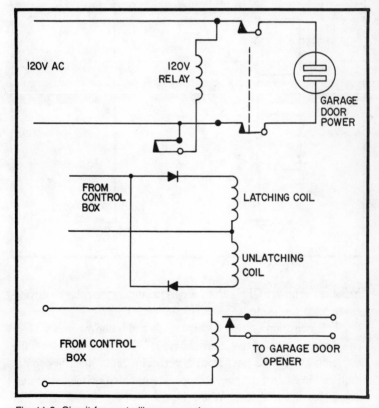

Fig. 11-8. Circuit for controlling garage door.

209

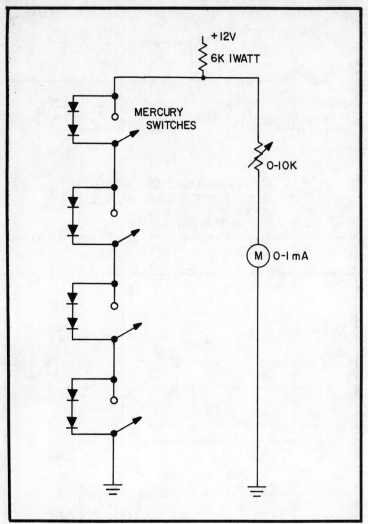

Fig. 11-9. Garage door position indicator.

Pressing the power OFF switch simply unlatches the relay removing power from the system.

The indicating system is very simple and is shown in Fig. 11-9. The garage door is of the overhead type. When the door opens, the panels one at a time pass from a vertical position to a horizontal position. On each of the panels there is a mercury switch connected across two silicon diodes. When the door is fully closed, all of the diodes are shorted out. Thus there is no voltage across the diode string and the voltmeter gives a zero indication. As the door opens the diodes are switched into the circuit two at a time causing the voltage across the string to rise. This causes the deflection of the voltmeter to increase.

A single snap action switch was installed at the bottom of the garage door as an additional security measure so that it was possible to tell when the garage door was fully closed, at a distance without looking at the meter.

INTERCOM CONTROL SYSTEM

The intercom control system is rather interesting. The complete schematic diagram is shown in Fig. 11-10. The section that

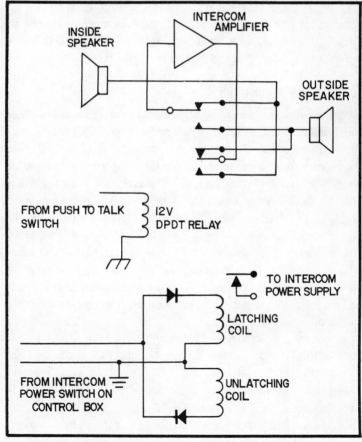

Fig. 11-10. Control of intercom system.

applies power to the amplifier is just like any other DC control circuit that uses a latching relay. When the on switch is closed momentarily, a positive signal travels down the control line and causes the relay to latch applying power to the amplifier. The push-to-talk relay is not energized at this time, so the intercom will be connected to transmit sound from the outside of the house to the inside. When the push-to-talk switch is closed, the push-to-talk relay will be energized connecting the amplifier and speakers so that sound will be transmitted from the inside of the house to the outside. Inasmuch as it is not desirable to be able to leave the system in this condition accidentally, no latching relays are used. The system can only be set for talking from the inside of the house to the outside by holding down the push-to-talk switch.

An indicator light was provided in the control box to indicate when the system was turned on, but this was found to be superfluous because there is always enough background noise audible in the speaker to indicate that the system is turned on.

OUTSIDE LIGHT CONTROL

In the home in which this particular system was installed, there are four lights on the front of the house, including one over the front door. It was desired to be able to turn the lights on full when someone visited, but to be able to dim the lights when they were left on merely for security or decoration. It was also desired to be able to control the lights from both the living room and bedroom control centers. The circuit selected for the purpose is shown in Fig. 11-11.

The main power to the string of lights is controlled by an AC relay with heavy contacts that will handle the load easily. This relay, K2, is operated by a latching relay K1. Thus when a positive signal is transmitted from the control box, relay K1 will latch energizing relay K2 and turn on the lights. Transmitting a negative signal will turn the lights off.

Dimming is accomplished by another AC relay K4. The contacts of this relay cut a silicon diode in and out of the circuit. The relay is controlled by the latching relay K3. When the diode is switched into the circuit it will cut the power furnished to the lights in half. This reduces the light intensity to less than half of full brilliance because the light bulbs are not as efficient at the reduced power.

The outside lights are not visible from the bedroom control box,

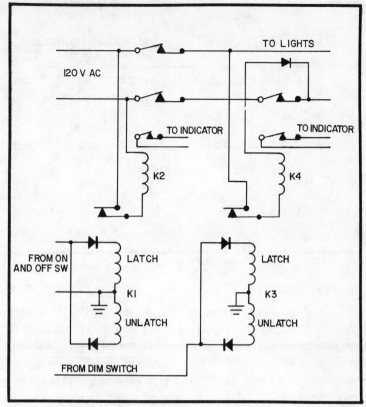

Fig. 11-11. Control of outside lights.

so it was felt necessary to add an indicator at least to the bedroom
control box. In order to prevent the lights from being left on during
daylight it was felt worthwhile to also include the same indicator on
the living room control box. The indicator signal is derived from
another pair of contacts on relay K2 as shown in Fig. 11-12. When
power is supplied to the lights lamp L1 will glow. In this particular
circuit, another lamp connected to relay K4 provides a signal to lamp
L2 that indicates when the lights were dimmed. This probably isn't
necessary, but was added so that the lights wouldn't be left at full
brilliance when they should be dimmed to save energy.

Of course, it is possible to add a dimmer circuit to provide
continuous control of the level of the lights. In fact this was tried, but
it complicated the control circuitry and was found to be unnecessary
because two levels of light were all that were needed.

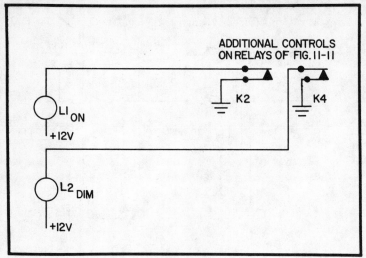

Fig. 11-12. Status indicator for outside lights.

DRAPERY CONTROL

The drapery control system shown in Fig. 11-13 is actually very simple, although many unsuccessful mechanisms were tried before the final successful system was built. Most of the mechanisms that were tried had some limitation or another with the result that operation wasn't reliable. The successful solution to the problem used an electric automotive radio antenna. This particular model was salvaged from a junkyard from a 1969 Buick, but new units are available for raising and lowering CB antennas. At first it was thought that it would be necessary to remove the antenna element before attaching the motor mechanism to the drapery cord. Actually, the antenna was left in place and merely secured to the drapery cord. The antenna rod increased the rigidity of the assembly, making it easier to move the cord in an upward direction.

The antenna raising motor operates from a 12-volt DC source, so the power to operate the motor could be carried directly from the control box. In this case, small wires were already installed so a simple DC power supply, shown in Fig. 11-14, was installed at the site. The drapery installation was such that the entire drive arrangement could be concealed.

The most important part of this installation and probably of any installation where mechanical motion is required was getting the drapery traverse rod and cord arrangement in good condition before adding the control device. Although it wasn't not noticeable when the draperies were operated by hand, the mechanism was actually anything but smooth. A little cleaning, lubrication, and adjusting of pulleys made a remarkable improvement.

The final arrangement operates very smoothly and is one of the most impressive features of the entire control system.

STEREO VOLUME CONTROL

The next row of switches controls the level of the stereo system. It was found that inasmuch as the turntable of the stereo system is located some distance from the speakers, it was impossible to adjust the volume to what would be a comfortable volume at the listening position. Furthermore, it was not desired to make any modifications to the stereo system itself. The system finally selected is shown in Fig. 11-15. Here a small reversible AC motor is used to drive a stereo attenuator.

The motor operates on 120 volts AC simply because a geared motor of this type was available for a couple of dollars at a surplus

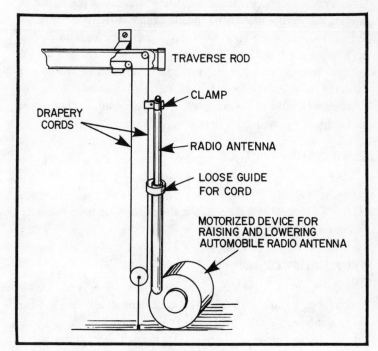

Fig. 11-13. Device for opening and closing draperies.

store. Actually any low-speed reversible motor can be used. Power is applied to the motor through two relays as shown in the figure. Inasmuch as the attenuator will only be used occasionally until the volume of the stereo signal reaches the desired level, there was no need for latching relays.

The mechanical connection between the motor and the attenuator is shown in Fig. 11-15. The biggest problem was that if the coupling between the motor and the attenuator was too tight, it was possible to destroy the attenuator by driving the motor after the attenuator had reached the limit of its travel. This was solved by simply loosening the mechanical coupling between the motor and the attenuator. Now, when the attenuator reaches the limit of its travel, the shaft will slip.

One of the most useful features of the attenuator was that it permitted reducing the volume during telephone conversations. Because of this, another muting system was made for another application. The arrangement is shown in Fig. 11-16. Here an attenuator is also used, but it has a fixed setting and no motor is used. Here a relay simply cuts the attenuator in and out of the circuit. A latching relay is used to control the main relay. When it is desired to cut the volume of the stereo a positive pulse is transmitted. This latches relay K1 and energized relay K2 which switches the attenuator into the circuit. Transmitting a negative pulse will unlatch the latching relay and switch the attenuator out of the circuit. Another relay K1 and the other half of the attenuator are not shown in the figure to avoid cluttering it.

OTHER CONTROL SWITCHES

All of the other control switches on the control box of this particular system are simple on-off switches that drive latching relays. They need no further explanation. They are used to control lights, air conditioners and fans.

THE AUXILIARY CONTROL BOX

As noted earlier in this particular system there is an auxiliary control box located in one of the bedrooms of the house. This particular box only has seven sets of control switches, because only a few of the functions need to be used at night. The first three sets of switches control the garage door, the intercom, and the outside

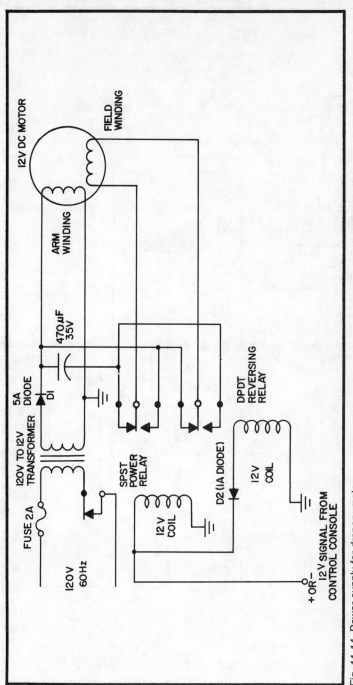

Fig. 11-14. Power supply for drapery motor.

217

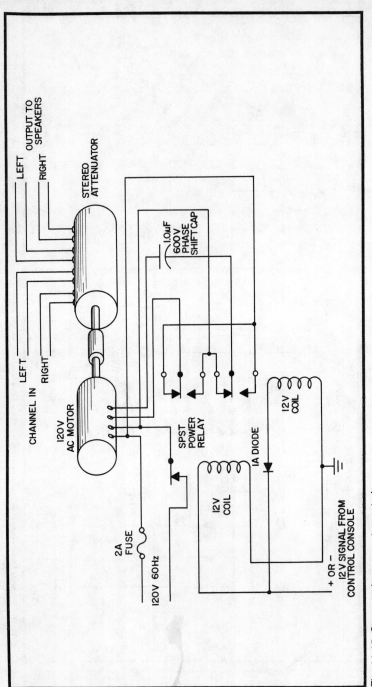

Fig. 11-15. Stereo volume remote control.

lights and are connected to the same control wires that are used by the main control box.

The only limitation of this system is that if the on switch on one control box is pressed at the same instant that the off control switch for the same function is pressed on the other box, there will be a short circuit path as shown in Fig. 11-17. Inasmuch as most of the time there is only one person operating this system, the likelihood of this happening is very slight. It has never been a problem in this particular application.

KEEPING TRACK OF WIRING

Most control systems are not finished in a day or even a week. The construction and debugging process is carried out over a long period. When the wiring is first started and everything is fresh in the

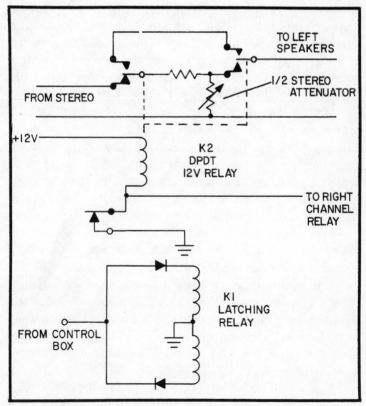

Fig. 11-16. Stereo muting control. Another DPDT relay and half an attenuator is used for the right channel.

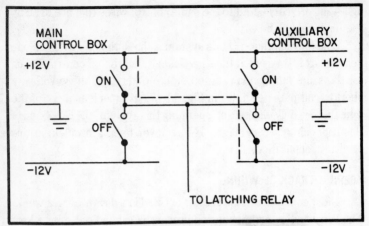

Fig. 11-17. Dashed lines show possible short circuit path.

constructor's mind, there seems to be no need to set up a system for keeping track of wiring. This is a shortsighted approach, because it is amazing how fast one can forget some of the details of the system if they are not properly documented. The small amount of time required to properly document a system will be more than repaid in the time that might otherwise be spent in ringing out unknown wiring.

At first it might seem that a properly used color code would take care of everything. Use of a color code for wiring is good and is encouraged, but it has its limitations. Often the wiring between the various parts of the system is made with multiple conductor cable that is purchased from a surplus house. Sometimes the colors used in one cable do not match those used in another cable and it is necessary to splice one color wire to a wire of a different color. There is nothing wrong with this, if it is properly documented. It is indeed frustrating to find that a red wire at one point is spliced to a wire of another color, but the exact color is forgotten.

It usually isn't necessary to make a complete wiring diagram of the entire system. In fact, such a diagram is apt to be confusing because usually only one part of the system is being checked at any particular time. The best approach is to make wiring lists with plenty of notations of the details. All terminal strips should be numbered to make circuit tracing easier.

Figure 11-18 shows part of the wiring list for the cable coming out of a control box of a system. Note that both the colors of the wires and the numbers on the terminal strips are recorded. It is easy

220

to identify the wires and terminals that are used for any control function.

In the particular system we have been describing in this chapter, the power supply for the main control box is located in the garage just the other side of a wall near the location of the control box. There were two reasons for doing this. First, and probably most important, it meant that the highest voltage at the control box was 12V DC. This made the box safe under all conditions. The second consideration was that although the power supply could have been made to fit into the control box, leaving it out left more room and as there are many connections the person working on the box can use all of the space he can get.

All of the control functions of this system were brought to a set of terminal strips mounted on a shelf in the garage. This called for a little more wiring that was actually necessary but made it very convenient to make changes or to troubleshoot the system. The set of terminal strips is shown in Fig. 11-19. The illustrations are self explanatory. The figure shows how easy it is to make measurements or to make changes on the system. New devices can be added and devices that have not been found useful can be removed easily and the circuits used to control something else.

KISS

The KISS formula mentioned early in this chapter was used throughout the system and it certainly paid off. Before the system was finally completed, many rather elaborate devices had been planned. In general, the more complex the circuit, the less likely it would ever be completed. When the system was first installed only the first three sets of switches on one control box were in operation, but the cabling for all of the other functions was installed and all of the switches were mounted on the control box. The system will finally

SWITCH	FUNCTION	TERM NO	COLOR
1A	GARAGE POWER OFF	1	RED
1B	GARAGE POWER ON	2	ORANGE
1C	GARAGE DOOR OPERATE	3	YELLOW
2A	INTERCOM ON	4	GREEN

Fig. 11-18. Sample wiring list.

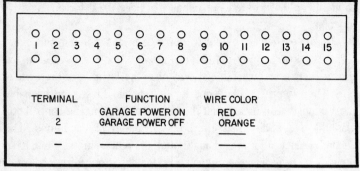

Fig. 11-19. Sample terminal strip identification and wiring list.

completed working evenings and weekends by building a component at a time and installing it. Because of the simple layout, it wasn't necessary to disturb anything that was already working while adding a new feature.

As of this writing the system has been in operation for over four years with very little trouble. One latching relay failed for no reason that could be found. The replacement has worked satisfactorily. A poorly soldered connection on one of the control switches caused erratic operation until it was found and corrected. The only other problem has been that some of the indicator lamps burned out. This situation was corrected by adding low value resistors in series with all of the indicator lamps. The brilliance was reduced, but none of the lamps have failed in three years.

In the particular home where this system was installed it was not convenient to lock the door between the house and the garage. Thus it was possible that anyone having a garage door radio transmitter of the proper frequency and modulation could gain entrance to the home. This, of course, wasn't likely, but it was decided to provide a check to see if it ever did happen. This was accomplished by connecting an electrically operated counter to the switch that closed when the garage door was fully closed. The counter was mounted on the front of the control box and could be reset mechanically. Thus the counter counted the number of times that the garage door was opened since it was last reset. By checking this counter after one returned from being away, it could be determined whether or not the garage door had been opened during the absence. The feature was probably not necessary, because there hasn't been such an intrusion in the four years that the system has been functioning.

12
Closed Loop Control Systems

All of the systems described so far in this book can be called open loop systems. These systems have no complete loop from the input through the output and back to the input again. As a result, they are really remote control systems rather than fully automatic control systems. With the exception of the time controlled systems, they all require a human operator.

The open loop system is easy to design and get running, and it is fully adequate for most applications. Generally the fact that someone has to initiate the control action is not a serious limitation. There are a few places, however, where a fully automatic system is desirable. An example might be control of an attic fan. We might want the fan to turn on whenever the temperature in the attic exceeds a certain value. Another example would be when we want a curtain to close when direct sunlight is shining into an area.

Fully automatic systems of this type are closed loop systems. To design and build a completely new closed loop control system is no simple engineering task. To make it work properly is even more of a job. Fortunately, many closed loop systems can be constructed from readily available components.

Figure 12-1 shows the functional elements of a closed loop control system. The first functional unit is some sort of sensor. This device senses whatever it is that we might wish to control. One of the sensors described in Chapter 2 will be adequate in most cases.

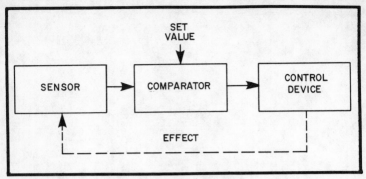

Fig. 12-1. Elements of a closed loop control system.

The next functional unit is a comparator. Here the signal from the sensor is compared with a set value of whatever we want to control. In some cases, such as a household thermostat, the sensor and comparator are contained in the same unit.

The output that we need to get from the comparator depends on what our control system will do. Often a simple on-off signal is all that is required.

The final element is the control device. This tells the end device of the system what to do. In the case of a furnace, for example, it will turn the furnace on or off.

Because of the fact that the loop is closed, it is capable of oscillation. This oscillation may range from "hunting" to wild swings of whatever is controlled. One of the objectives of system design is to minimize this oscillation.

LIQUID LEVEL CONTROL

Two different liquid level controls will point out the differences between various types of control systems. The first example, shown in Fig. 12-2 is an automatic sump pump which is intended to pump water out of a leaking basement.

The requirements of this system are simple. We only want to turn the pump on when there is water in the basement and turn it off when the water has been pumped out. We would never want to reverse the procedure and pump water into the basement. All we need for a control signal is some sort of on-off signal. It is apparent that some sort of switch could be used for this purpose.

The first task is to convert the presence of water in the basement to an electrical signal. This can be done with a float arrange-

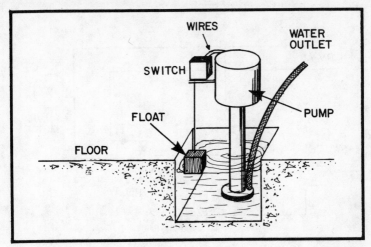

Fig. 12-2. Automatic sump pump.

ment shown in Fig. 12-3. Here a float is connected to the bottom of a rod and a snap-action switch is connected to the top of the rod. The assembly is located so that the float is in a small sump that will collect any water that leaks into the basement. When the water collects, the

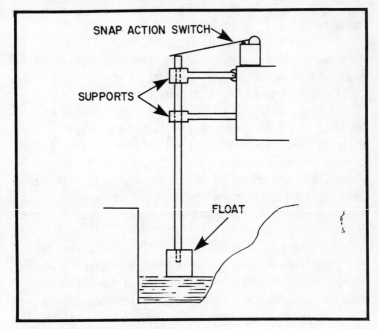

Fig. 12-3. Details of float switch.

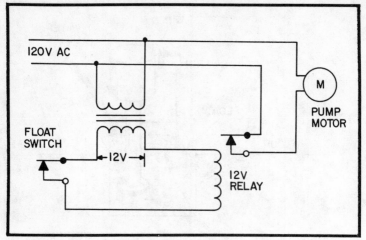

Fig. 12-4. Circuit of automatic sump pump.

float will rise closing the snap action switch and starting the pump. When the water is pumped out, the float will drop, opening the circuit and shutting off the pump. The complete circuit is shown in Fig. 12-4. The actual control device is a relay with a 12-volt coil and contacts that will safely carry the operating voltage and current of the pump motor.

The shaft between the motor and the pump, and the rod that actuates the snap-action switch are long enough so that none of the electrical components will ever be under water.

Because of the nature of the snap-action switch, this arrangement will not hunt or oscillate. Before we discuss this, let's look at what might make the system oscillate. Consider the situation shown in Fig. 12-5. Here the water is right at the level where the float will close the switch and start to pump. If the water level rises, ever so slightly, the pump will start. If the water level drops ever so slightly the pump will stop. It is possible that even the turbulence in the water caused by the pump will be enough to make the system oscillate—on off, on off.

One way to avoid oscillation is to avoid the critical water level of Fig. 12-5. What we want is a situation where the pump will start when the water reaches one level and then will stop when the water reaches a lower level as shown in Fig. 12-6. Here the pump will start when the water reaches the higher level and will stop when it reaches a lower level. It won't start again until the water reaches the

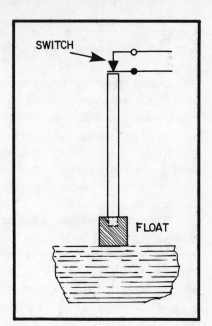

Fig. 12-5. Critical water level.

higher starting level. Usually this will take quite a bit of time, so the system won't oscillate.

One way to get the difference between the starting and stopping levels is to use a snap-action switch with some hysteresis; that

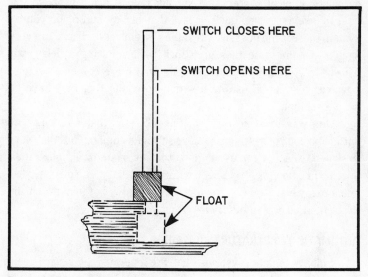

Fig. 12-6. Avoiding the critical water level.

is one position of the switch lever will close the switch, but the switch won't open until the lever is some distance below the starting level.

Another way to avoid oscillation in a system like this is to put a timer on the motor so that it will continue to run after the sensor switch has opened. With this arrangement, the pump motor will start when the water reaches some given level. When enough water is pumped out so that the water is back to the starting level, the sensing switch will open. Normally the motor would stop at this point, but the timer will keep it pumping until after the water reaches a lower level.

This system can have some· additional features added. For example, suppose that the pump had enough capacity to keep up with any normal leaking of water into the basement. Under these conditions the pump would only have to run for, say, ten minutes to bring the water to the stop level. A timing circuit could be connected to the motor supply in such a way as to sound an alarm if the pump operated continuously for more than 20 minutes. This would alert the homeowner to any potentially disasterous situation, such as a broken water pipe, where the pump couldn't keep ahead of the rising water.

The automatic sump pump described here is a very simple automatic control system and shouldn't pose many problems. It does, however, point out the fact that in any automatic control system there is the possibility that the system will oscillate. In a very simple system some measure, such as introducing a delay, will usually solve the oscillation problem. In a more complex system, however, oscillation can be a serious problem that may have no simple solution.

One way to avoid many of the problems that go along with automatic control systems is to use salvaged components that were designed originally for use in an automatic system, although their specific function may have been different. Usually, the designer has spent many hours on features that hardly show, but minimize many problems that would otherwise occur.

AUTOMATIC TEMPERATURE CONTROL

Fully automatic systems that are operated by a change in temperature of the surrounding air are probably the easiest to build

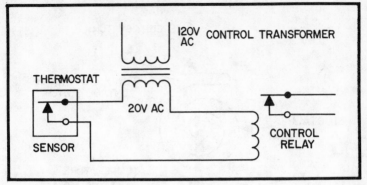

Fig. 12-7. Elements of a temperature control system.

of all automatic control systems. The reason is that all of the components needed to build such a system are available at hardware and plumbing supply stores. The only problem is to lash them together properly.

Figure 12-7 shows the basic system. The sensor and comparator are simply an ordinary household thermostat of the type used to control a furnace in a home. The control signal is usually a 20-volt AC signal. The thermostat contains a mercury switch that closes when the temperature falls below a certain value and opens when the temperature reaches a higher value.

Motor control relays of the type used in furnaces will operate on the 20-volt control signal. A relay of this type can be used as a control device to operate almost anything that works on an electric power line.

A system like that in Fig. 12-7 can be used to control fans, air conditioners, and of course furnaces.

Figure 12-8 shows a rough sketch of an ordinary household thermostat. The temperature sensing element is a bimetallic spiral. The inner end of the spiral is anchored to the frame of the thermostat and a mercury switch is attached to the other end. As the temperature of the surrounding air rises, the bimetallic spiral will expand causing it to rotate. This rotation will cause the mercury switch to tilt closing the circuit.

Many thermostats have an interesting arrangement that is designed to prevent the system from oscillating. Here's how it works. When the temperature of the surrounding air falls below a certain value, the mercury switch will tilt, turning on the furnace.

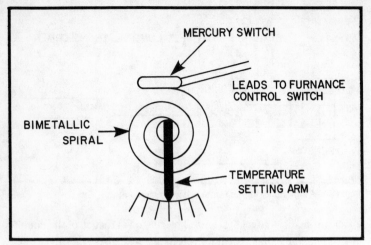

Fig. 12-8. Basic elements of a household thermostat.

The way things are laid out in many homes, by the time the spiral in the thermostat gets hot enough to turn off the furnace, the rest of the house is too hot.

To prevent this, when the mercury switch is closed, a small current flows either through the bimetallic spiral, or through a heater that will heat the spiral. This extra heat causes the mercury switch to turn off before the thermostat reaches its final temperature. In this way, the thermostat anticipates when the room air will reach the desired temperature. This small heating current will prevent the system from cycling rapidly. This feature can also be used to stabilize any control system in which a household thermostat is used as a sensor.

AN ATTIC FAN CONTROL SYSTEM

In many climates, a fan which will exhaust hot air from an attic will make a home much more comfortable. Figure 12-9 shows such a system. The sensor is a household thermostat and the control device is a relay designed for controlling a furnace motor. Power is supplied from a 24-volt control transformer intended for use in a furnace system. All three components are readily available and the system can be installed in a few hours.

The thermostat can be set to operate whenever the temperature in the attic exceeds some value. Most household thermostats have a range of 50 to 100 pF which is adequate for such an application. Naturally it wouldn't be advisable to use such a system in the wintertime. Then, any hot air in the attic would be an asset. It would reduce the fuel bill.

The simple approach is to turn off the control system during the winter months when the home is being heated. For the buff who wants the system to require no attention at all, even this function can be automated. Fig. 12-10 shows the circuit. Here two thermostats are used in series. One thermostat is located in the attic as in Fig. 12-9. The other is located outside the house. Before the fan can turn on, both thermostats must be on.

Thus, for example, the system can be set so the attic fan will only turn on when the attic temperature exceeds 90 degrees and the outside temperature exceeds 70 degrees.

AUTOMATIC LAWN SPRINKLER

Figure 12-11 shows a fully automatic system that will turn on a lawn sprinkler when the earth is dry and turn it off after a sufficient amount of moisture has been absorbed.

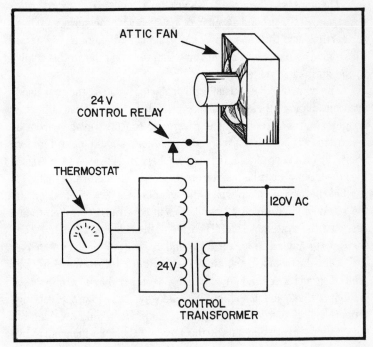

Fig. 12-9. Automatic attic fan.

231

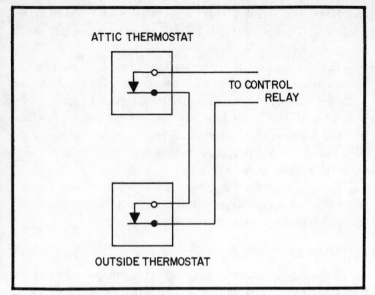

Fig. 12-10. Attic fan that takes outside temperature into consideration.

The sensor is basically an ohmmeter circuit that measures the resistance of the ground a couple of inches below the surface. When the earth is dry, the sensor provides a signal to the control system. This signal actuates a solenoid valve that turns on the water supply to the sprinkling system.

The control valve is usually no problem. It can be the type used in an automatic clothes washer to control water flow. It can either be bought new as a replacement part or it can be salvaged from a discarded washer. Preferably the solenoid should be rated for operating on a low voltage. In many washers this is done. In this case the power supply can be taken from the control transformer that was used in the washer from which the solenoid valve was obtained. If the solenoid operates at 120-volts, great care must be taken to be sure that the system is safe. The case of the solenoid valve must be securely grounded to an earth ground.

The sensor of the system is the moisture sensor made from a printed circuit board and described in Chapter 2. The sensor is easy enough to make and install, but it isn't always easy to get it working properly. The secret is finding where to place it and how deep to bury it. This in turn depends on the nature of the soil which may vary considerably from one location to another.

The best approach is to find the correct location of the sensor by trial and error. This means arbitrarily placing it in some location and watching the system at work. If the system turns the sprinkler off before the soil has absorbed enough moisture, the situation may be improved by changing either the depth or the location of the sensor.

There are many additional features that can be added to this system. For example, some plants do better if they are not watered while the sun is shining brightly. A solar cell may connect in a circuit where the output will vary with the amount of sunlight. When the sunlight is greater than some arbitrary value, the sprinkler will not work.

If it is desired for the system to only operate at night, a timer can be added. In this circuit both the timer and the sensor must supply a complete path for the current to the valve before it will open. The timer can be set so that its contacts will only be closed at night. Not only will this prevent the plants from being watered during periods of bright sunlight, but it will prevent the sprinkler from turning on during the day when people may be within its range.

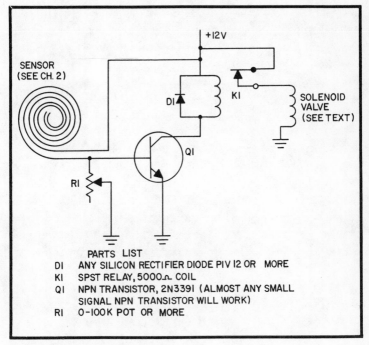

PARTS LIST
D1   ANY SILICON RECTIFIER DIODE PIV 12 OR MORE
K1   SPST RELAY, 5000Ω COIL
Q1   NPN TRANSISTOR, 2N3391 (ALMOST ANY SMALL
     SIGNAL NPN TRANSISTOR WILL WORK)
R1   0-100K POT OR MORE

Fig. 12-11. Automatic lawn sprinkler.

OTHER FULLY AUTOMATIC SYSTEMS

In this chapter we have given examples of a few automatic control systems. Our purpose has been primarily to point out the problems that arise in connection with such systems and to make suggestions as to how to solve them. We have intentionally avoided any mathematical discussion of the subject. Not that the reader would not understand it, but usually there isn't enough quantitative information available for solving the equations that arise in connection with the design of a system. The approach should therefore be a pragmatic one in which the system is altered until it works properly.

There is no reason why other control functions cannot be fully automated, but usually it is not necessary and except for a few rare instances, it is not worth the trouble. If it is felt that some control function should be automated, the best approach is to see if there isn't some way to modify components that were originally designed for use in automatic systems. These components usually include features that solve many of the problems, such as oscillation, that are typical of automatic systems.

13
Computer Based Control Systems

The ultimate in control systems is the system that has a digital computer as its "brain." Until a few years ago, it was unthinkable that a computer could be incorporated into anything but the most sophisticated industrial or government control system. It certainly couldn't be used in a home system. A digital computer with any capability at all, probably wouldn't even fit into the average home, let alone be a part of a control system.

The advent of the microprocessor which places the entire central processing unit of a digital computer in a single integrated circuit has made the computer not only small, but inexpensive. Thousands of small computers, normally called microcomputers, are in use in homes, businesses, and in the dreams of hobbyists. Adapting such computers for use with a home control system is really quite simple. The biggest problem is interfacing the computer with the other elements of the control system, which is the subject of the following chapter.

All of the control systems described in this book operate at a snail's pace compared to the speed of the slowest microcomputer. This, in practical terms, means that a microcomputer can not only handle all of the decisions required in a home control system, but it can do it in its spare time and still be available for all the other normal uses. For example, a computer can be used for computations, data processing, or even a game, and at the same time handle

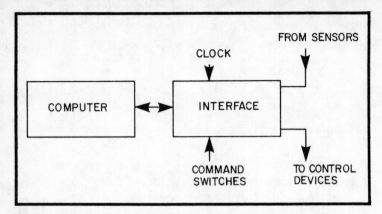

Fig. 13-1. Elements of a computer based control system.

everything required by the average home control system. In fact, the amount of time required by the control system is so small that the user who might be playing a game with the computer would never recognize the fact that the computer might have interrupted the game and sent a few signals to the control system.

The biggest advantage of using a computer to handle a control system is that the computer can take a very large number of things into consideration before making a control decision. This doesn't require increasing the complexity of the control system. All of the complexity can be incorporated in the computer program. The actual control system might well be simpler than the conventional systems described in earlier chapters of this book.

Figure 13-1 shows a block diagram of a computer based control system. Here all of the sensors and all of the command switches of the control system are connected through the interface to the microcomputer. Similarly, all of the control signals come from the microcomputer, through the interface, to the control devices. We have shown a clock as one of the inputs to the system, but this really isn't necessary. The computer itself, once it has been properly programmed, can compute the time of day. In other words, it can have its own internal clock. In fact it can not only keep time, but if desired it can correct itself regularly.

The arrangement of Fig. 13-1 can be used as a fully automatic control system, or as a remote control system or both. When used as an automatic system, the computer will periodically examine the state of the various sensors, compare these states with things stored in its memory, and when required by the program, make a control decision and send a control signal to one or more of the control devices.

When the system is used as a remote control system, it is controlled by signals from the outside. These signals may be commands typed into the keyboard of the computer, or they may originate from dedicated command switches, just like those used in a conventional remote control system. The advantage of using a computer in a remote control system is that it can take many things into consideration before sending out the control signal.

CONTROL PROGRAMS

The exact programming of a computer for operating either an automatic or a remote control system will differ from one microcomputer to another depending on the instruction set of the basic processor and the type of language used to program it. If some form of basic language is incorporated in the computer, programming is extremely simple. In fact, most of the programming required for operation of a home control system is so simple that it can be done easily using machine language.

The operations required for operating a control system are listed in Table 13-1. Of course, the program can be made as sophisticated as the user wishes.

Many of the sensors described in the systems of this book have the advantage that they are simply switch closures that indicate that a given situation is either true or not true. The light level sensors simply decide whether or not it is dark in a certain area. The door position sensors simply tell whether a door is opened or closed. Any sensor of this type requires only one bit of an 8-bit input bus to get

Table 13-1. Control Systems Operational Requirements.

```
  I. SENSE CONDITION OF SOMETHING
        a. SENSOR
        b. CONTROL SWITCH
        c. PROGRAM
        d. TIME OF DAY
  2. MAKE DECISION
        IS A CHANGE NECESSARY
  3. INITIATE CONTROL
        SEND A CONTROL SIGNAL
  4. CHECK TO SEE THAT CONTROL OPERATION
        WAS PERFORMER (OPTIONAL)
```

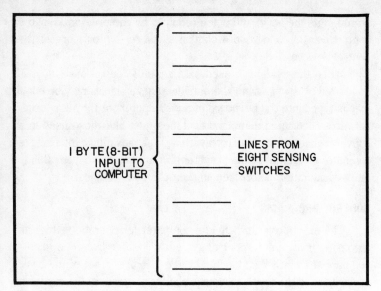

Fig. 13-2. Using sensor switches to form a byte of data.

information into the computer. Thus very little of the computer capability is required. In fact, the control system is so simple compared with the capability of the computer that it is often practical to use a very simple, but inefficient way to get information in and out of the computer. This results in very little hardware being added to either the computer or the control system.

Figure 13-2 shows an example of where a single 8-bit word is used to tell the computer the state of 8 different on-off sensors. Here the computer reads the 8-bit word out of the interface which we discuss in the following chapter. Each bit of this word tells the state of one of the sensors or command switches in the system. The 8-bit word, or byte, is fed into the accumulator of the computer where it is compared with a reference word in the memory. This comparison may be made either by a comparison instruction, or by a masking operation that will tell which, if any, of the bits in the word have changed with respect to the reference word stored in the memory. Suppose, for example, that the third bit of the word were connected to the command switch that told the computer to open the front door. The computer would detect the fact that this had changed and would issue the command signal needed to operate the door opener.

Of course, a computer isn't needed for such a simple operation. It might still be advantageous to use the computer even for such a simple control problem because of the many things that could be made to enter into the control decision. For example, the computer might not respond immediately to the command, but check some other things first. There might be a pressure switch in a mat inside the door where it would swing when opened. The computer could check the state of this switch before sending out the command to open the door. If someone were standing where he might be struck by the opening door, the computer would not issue the instruction. In fact, it might even go so far as to activate a warning buzzer if anyone were standing close to the door.

STATUS DISPLAYS

In the systems described in earlier chapters, indicators were provided to show the status of anything that might not be obvious to anyone operating the control system. Such indicators may still be used with a computer based system. They are either connected directly to the sensors, or are driven by signals from the computer. The display of the computer may also be used to indicate the status of the various sensors in the control system. This may be a composite display, or a display of individual conditions.

Figure 13-3 shows an arrangement where all of the sensors that indicate—things such as open doors or windows, appliances running, or anything else that might influence the security of the home—are tied to the same 8-bit input to the computer. When the command that indicates that the homeowner is going to bed or leaving the premises is typed into the keyboard of the computer, the status of this byte is automatically checked against a word in the computer memory. If the condition isn't secure, the computer will detect condition and will display the insecure condition. For example, it might display the message, "BACK DOOR OPEN."

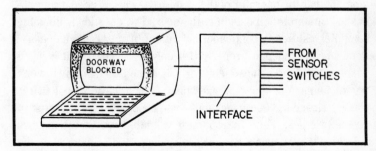

Fig. 13-3. Using the computer display.

The display of the computer can be used to give any desired indication of the status of the control system. If desired, it can include a number of self checking features where the display will tell the probable cause when any portion of the system fails to function properly. In the example we gave above of the computer checking to see if anyone was in a position where they might be hit by an opening door, the display might indicate, "DOORWAY BLOCKED."

These uses of the computer display are almost unlimited. The status of everything that has a sensor is available to the computer. All that has to be done is to write a program that will develop the desired message from the states of the various sensors.

CONTINUOUS SENSORS

Obviously, the easiest type of sensor to interface with a computer is the one that gives an on-off indication. Unfortunately a control system occasionally needs a sensor that will give more information about the state of something that can be contained in one bit of information. In the extreme case, the output of the sensor is an analog signal. The computer can't handle analog signals, so these analog signals must be converted to digital form before they are applied to the computer. If a very detailed signal is required, an analog to digital converter may indeed be required. Usually, with a little ingenuity, a way can be found to get all of the information that is really necessary without completely digitizing the signal from the sensor.

Figure 13-4 shows an analog display of the state of a door using the sensor described in Chapter 2. This is indeed an analog signal and it certainly can't be fed to the computer. The reason that we used this type of sensor in the first place is that we felt that a continuous indication of the door position was necessary. Let's assume for the moment that the operator of the system does really need a continuous indication of the position of the door. The question is does the computer also need a continuous indication? In most cases it doesn't. It might well function properly if it only has signals that indicate whether the door is fully opened or fully closed. In this way it won't try to open a door that is already opened or to close a door that is already closed. These indications can be obtained from two snap action switches. Then the operator will still have the continuous indication that he wants and the analog signal will be ignored by the computer which doesn't need it anyway.

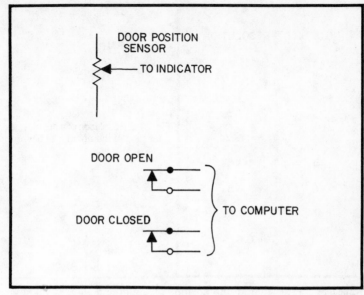

Fig. 13-4. Continuous sensing for indication. Switch sensing for a computer.

As we have seen, it is much more convenient to use a simple on-off sensor with computer based systems. Nevertheless, there will be cases where we need much more than a simple on-off signal. We will discuss various approaches to this aspect of the problem in the following paragraphs.

As we have noted, an analog signal is a continuous function of somthing. It varies smoothly from one value to another. A digital signal is not. Actually there is more to the situation than this. Although an analog signal is continuous, we can't trust it down to extremely close resolution. Any analog sensor will have some error, some nonlinearity, and some noise. Thus although the signal may be continuous, we really don't need a truly continuous signal because it will always have some error anyway.

Now that we know that we can live with a digital indication from a sensor, the next thing we have to decide is how many bits we need in a digital signal to give us the resolution. Figure 13-5 shows the resolution obtainable with digital signals having different numbers of bits. If one switch, say at the middle of the range of travel of a door were used, we would know the value signal to within 50% of its possible values. We would know whether the door was closer to being fully opened or fully closed. At the other extreme, if we had an

NO. OF BITS	RESOLUTION
1	50 (%
2	25 %
3	13 %
4	6 %
5	3 %
6	2 %
7	1 %
8	0.4 %

Fig. 13-5. Resolution of a quantity as a function of the number of bits.

8-bit signal representing the position of something we would know the position to within one part in 256 or to about four tenths of one percent. It is rather obvious that in most control systems what we need is somewhere between these two extremes.

The best approach to getting a digital input that is a measure of the position or state of something is to use a series of switches. The switches in Fig. 13-6 might be mercury switches that open as a garage door opens. As the door starts to open the top switch opens, first, then the next one, and so on. These switches will provide indications of the position of the door. They are not the usual digital signals that we think of in connection with computers because they do not represent a binary number that is a quantitive measure of the position of the door. Nevertheless, these signals are adequate for most systems and the transducers are cheaper that true digital sensors.

Figure 13-7 shows the four-bit digital output from the sensor of Fig. 13-6 together with its meaning in terms of the position of a garage door. Also shown is the decimal equivalent of the digital signal considering it to be a true binary number. The fact that the output of the sensor isn't efficient can be seen from the fact that a four-bit digital signal is used to give us five different values of the door position whereas in a true binary system four bits would give us 16 different values. That really isn't important. The important thing is

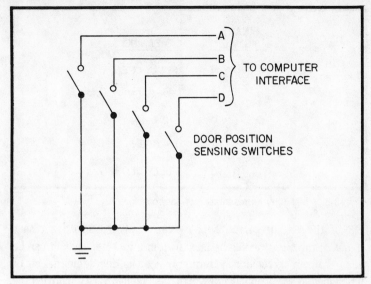

Fig. 13-6. Using position sensing switches to generate a digital word.

that the output of a sensor circuit such as that of Fig. 13-6 can be used quite well with a computer based system.

The way the output of the sensor is handled by the computer depends on the whim of the programmer. The signals can be applied directly to the data bus, or input, of the computer and the computer programmed to recognize the meaning of each of the four-bit words. The only real disadvantage of this arrangement is that it means that we tie up four lines of an 8-bit input bus of the computer to sense the condition of a signal sensor.

SIGNAL FROM FIG. 13-6				MEANING	DECIMAL NUMBER
D	C	B	A		
0	0	0	0	COMPLETELY OPEN	0
0	0	0	1	3/4 OPEN	1
0	0	1	1	1/2 OPEN	3
0	1	1	1	1/4 OPEN	7
1	1	1	1	FULLY CLOSED	15

Fig. 13-7. Meaning of signals from Fig. 13-6.

BINARY SIGNAL			MEANING
O	O	O	FULLY OPEN
O	O	I	3/4 OPEN
O	I	O	1/2 OPEN
O	I	I	1/4 OPEN
I	O	O	FULLY CLOSED

Fig. 13-8. True binary representation of door position.

If the output of the sensor were converted into a binary number, the signal would be as shown in Fig. 13-8. Here it can be seen that our inefficient system only cost us one additional bit or input line, so it is probably not worth the trouble to try to convert the sensor output into true binary form.

There is another trick that can be used to get a digital signal with a few bits that will give us all of the information we really need about a quantity that can vary over a wide range. That is to concern ourselves only with the values that are of extreme interest and to ignore other values. For example, suppose that we needed to sense the temperature in a room. Theoretically, the temperature could vary more than 100 degrees during the year. However, the only reason that we might wish to sense the temperature closely would be to control it and hold it to within a few degrees of some desired value. If the temperature were to fall below some arbitrary value, say 68 degrees, we would turn the heater full on. Therefore, we really don't need to know the exact temperature, just that it is below 68 degrees. Similarly, if the temperature exceeded some value, say 72 degrees, the heater would be turned completely off. Again we don't need to know the exact temperature. Now we have reduced the problem to where we want to know what the temperature is when it is between 68 and 72 degrees. If we had a separate thermostat element set to each of the four temperature values in this range we could know the temperature to within ½ degree. That surely should be adequate for any modest control system.

Thus although the sensors described do not provide an output that is a true binary representation of the quantity being sensed, they do give an output that can be handled by a computer and the resolution is adequate for most situations.

If the need is for a digital signal derived from the output of an analog sensor with a wide range and high resolution a true analog to digital converter would be required. Such converters are beyond the scope of this book.

GETTING THE COMPUTER'S ATTENTION

So far we have discussed the various signals that we might apply to the computer input from sensors and control switches. We haven't addressed the problem of getting the computer to pay any attention to them. There are two general ways in which we can get information into the computer. The first is by a polling routine where the computer periodically takes a look at the signals from the sensors and switches and then decides whether or not any action is necessary. This polling doesn't have to be done very often in terms of computer speed. In fact, it is probably better not to sample too frequently to be sure that the data has a chance to settle. We will say more about this in the next chapter. If the data is sampled once a second, the operator of the control system will not even notice the delay. In a computer having a clock rate of 1 MHz or faster, this allows plenty of time for other functions.

The other way to get information into a computer is to use the interrupt capability of the computer. When this is done, the computer will ignore all of the sensor signals until one or more of them change. The computer can be used for any other desired purpose while nothing is happening in the control system.

STORING THE PROGRAMS

One of the problems that must be faced sooner or later in using a microcomputer with a control system is that of properly storing the programs in the computer. Most hobby type computers are equipped with RAM or read-write memory for storage. There are addressable locations for ROM or read only memory, but programming such memories is often an expensive task that should be only undertaken after the program is completely debugged. The computer can also be equipped with a cassette recorder so that the program can be put back into the memory if the memory is erased for any reason such as a power failure.

RAM is absolutely the best way to store the program until it is completely debugged. Even the best programs are often found to have bugs after they are put into actual use. The debugging process must include letting the system perform all of its control functions under all possible conditions. It is very easy to forget one or more sets of conditions that may exist and that the program might not be able to handle. Only after every possible condition has been investigated is it safe to commit the program to some more permanent form of memory. The type of permanent memory used depends on what the user expects of the system and what he can afford.

One trick in programming a computer to handle control functions is to make liberal use of the NO OP (no operation required) instruction. This instruction does nothing but occupy space in the memory. By interspersing NO OP's liberally throughout the program, space will be provided for inserting steps that were forgotten during the programming.

Where the electrical power supply is reliable, it is feasible to use RAM for more or less permanent storage of the program. When this is done, provision must be made for disabling the computer if its memory is by chance erased. One way of doing this is shown in Fig. 13-9. The circuit shown in this figure is a power failure switch. When the system is set up, the SCR is fired by pushing the momentary contact switch S1. Once this has been done, the SCR will conduct and will continue to conduct until the power fails. The SCR energizes relay K1 which in turn energizes the circuits that allow the computer to send signals out to the control system.

If there is a power failure, the SCR will stop conducting, relay K1 will drop out, and the computer can no longer send signals to the control system. The trick is to make relay K1 drop out before the computer has a chance to do anything.

After power has been restored, relay K1 will remain open preventing the computer from sending any meaningless information to the control system. After the lost program has been put back in the memory from a cassette tape, switch S1 can be closed momentarily. This will allow the computer to again address the control system.

If some protective measure like this isn't taken, the computer may well send erroneous signals to the control system that will result in inconvenience or even danger. Anything that is normally controlled by the computer may behave very erratically.

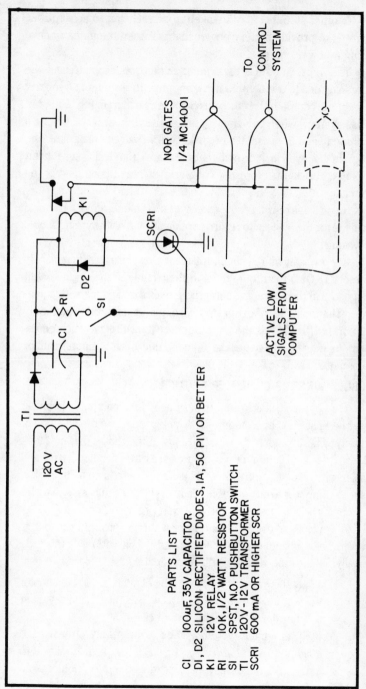

PARTS LIST
C1   100μF, 35V CAPACITOR
D1, D2 SILICON RECTIFIER DIODES, 1A, 50 PIV OR BETTER
K1   12V RELAY
R1   10K, 1/2 WATT RESISTOR
S1   SPST, N.O. PUSHBUTTON SWITCH
T1   120V - 12V TRANSFORMER
SCR1 600 mA OR HIGHER SCR

Fig. 13-9. Power failure protection circuit.

247

Another step that may be taken to protect RAM against power failure is to provide battery power that will keep the memory alive during a power outage.

The more ambitious experimenter may wish to automate the refreshing of a RAM memory in a computer. In the event of a power outage, the computer is removed from the circuit as in Fig. 13-9, but in this system, when power is restored, the computer will go into a read mode and the cassette recorder will reload the program. The program can even include self-check features that will assure that it is loaded correctly before the computer can take command of the system.

Regardless of what protection system is used, provision should be made for the system to return to manual control if anything should go wrong.

It is probably more economical to store the program in read only memory (ROM) than to go to all of the trouble of the arrangement suggested in the preceding paragraph. This can be done with equipment that can burn programs into programmable read only memories (PROM). If the experimenter doesn't have such equipment, he may be able to get the job done through a computer club or a computer store.

SUGGESTIONS FOR COMPUTER CONTROL

One of the feelings often experienced by a computer hobbyist after he finally has his computer working properly is the question of exactly what he should do with it. Many of the things that first look like an ideal application for a computer are often handled better and easier with a conventional control system. There is no point at all in using a computer to do something that can be done fully as well with a momentary contact switch and a latching relay.

The places to use a computer are where a conventional control system cannot handle the job conveniently. Applications of this type usually fall into the following three categories:

1. Many things enter into the control decision. For example, the outputs of many sensors as well as the time of day must be considered in making the decision.
2. Fully automatic systems. These systems are often hard to build and even harder to stabilize. It is usually easier to make the necessary modifications and refinements in software than in hardware.

3. Applications where the versatile display of the computer is necessary. Later we will describe an idiot-proof system that can be operated by anyone with a minimum amount of training in its operation.

A SOUND LEVEL CONTROL SYSTEM

The level of sound required from a TV set, radio or stereo depends on the ambient level of sound in the listening area. If the ambient sound level is low, not much volume is required for comfortable listening. If the ambient sound level in a room is high, it is often necessary to turn up the volume. On the other hand, when some other sound should take priority it may be necessary to turn down the volume.

A computer based control system can take all of these things into account when setting the volume of a stereo, for example. Figure 13-10, shows a stereo system with the record player itself located some distance away from the listener. In a conventional remote control system, a remote volume control could be located close to the listener so that the volume could be set to a comfortable level. With a computer based control system many of the volume

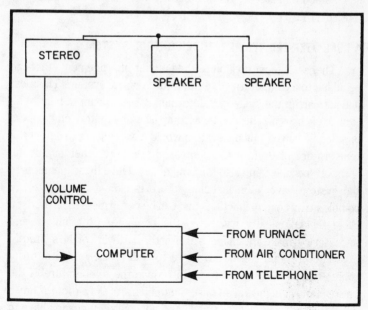

Fig. 13-10. Computer control of stereo volume.

adjustments could be made automatically. The system has a sensor in a nearby utility room where the furnace, which tends to be noisy, is located, a sensor on the air conditioner, and a sensor on the telephone. The computer can take the signals from all of these sensors and decide the proper sound level for the stereo.

In operation, the user sets the volume of the stereo to a comfortable level. When the furnace or air conditioner turns on, the system will automatically increase the sound level just enough to overcome the effect of the additional background noise. On the other hand, if the telephone is answered, the system will mute the sound from the stereo enough so that it will not interfere with the phone conversation.

The amount by which the computer must increase or decrease the sound level is determined by experiment and then programmed into the computer memory.

If it is desired, the system can also have an input corresponding to the time of day so that the stereo will not be played too loudly at night when people may be sleeping close by. This timing signal can come from either an external clock, or from a timing program in the computer. Of course, there is no reason why such a system should be confined to a stereo. It could equally well handle the output of a radio or TV.

A COMPUTERIZED TEMPERATURE CONTROL SYSTEM

There is almost no limit to the number of refinements that can be added to a computerized temperature control system. The most difficult part of the system is determining the amount of heat to be sent to each area. This can be accomplished by motorized dampers or solenoid valves. Such a system would have sensors to detect the temperature of each of the controlled areas, whether or not the areas are occupied and also the time of day. Once these elements of the system have been installed, there is almost no limit to the complexity of control functions that can be performed. From here on all of the complexity will be in the computer program. It isn't necessary to make any changes at all in the hardware of the system.

The program can be arranged to reduce the temperature of rooms that aren't occupied and to reduce the temperature of the entire home at night. It can raise the temperature of a room at times when the room is most apt to be used. Of course, it doesn't have to follow the same program every day. For example, it can raise the temperature later in the morning on weekends than on weekdays if this should be desired. It can even keep track of when heat is furnished to various areas of the home and the cost of heating these areas. The system can be programmed to warn when the heating bill is reaching an abnormally high level.

The many things that this system can do illustrate the principal advantage of using a computer in a control system. Expressed briefly, just about all of the complexity of a computer controlled system is in the program. This makes it possible to add all sorts of features without building any more hardware.

COMPUTER CONTROL OF MANY THINGS

Another outstanding feature of the computer controlled system is that any number of control functions can be added without taxing the capability of the computer. In the worst case all that will be required is some additional memory for the computer. If a computer system is once installed to control any function, additional devices can be controlled by installing the controls and sensors and reprogramming the computer. A single computer can handle everything.

14

Interfacing the Control
System with a Computer

Most microcomputers are very well suited for handling input signals from keyboards and audio tapes and for providing outputs to videodisplays, seven-segment displays and printers. They are not well suited to dealing with input signals from the real world and providing output signals that are capable of operating control devices. Before such a computer can be connected to a control system an interface must be provided that will convert the input signals to a form that can be handled by the microcomputer and convert the output signals from the microcomputer to a form that can operate control devices.

Figure 14-1 shows a block diagram of a microcomputer used as a part of a control system together with a list of the types of operations that must be performed by the interface. We will discuss these functions one at a time and suggest circuits that can be used to accomplish the various operations. There is usually more than one way to handle the signals in a computer based system and no pretense is made of covering all or even the optimum approaches to the interface problem. The circuits presented do, however, have the advantage of simplicity and will minimize the problem of mating a microcomputer to a control system.

LEVEL SHIFTING

As we have noted earlier, it is best if the sensors and command switches of the control system furnish on-off signals that are compat-

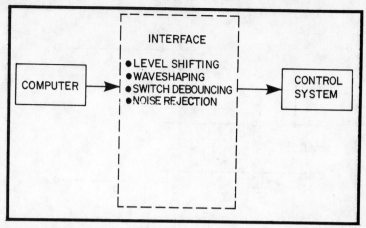

Fig. 14-1. Interface operations.

ible with the signal requirements of the microcomputer. Usually this means that the low signal must be between 0V and 0.8V and that the high signal must be between 2.4V and 5V. If the input signals do not fall in this range, their levels must be shifted. Signals that go negative or exceed +5V may damage the computer circuitry.

Figure 14-2a shows a typical level shifting problem. When the push button switch in the circuit is closed the relay will be energized and will apply +12V to the input to the computer unless we do something to change the situation. The solution in this case is rather simple. All we have to do is to add the voltage divider shown in Fig. 14-2b consisting of resistors R1 and R2 to divide the 12V down to 5V.

The design of the divider is rather simple. The first requirement is that the bottom resistor R2 be quite small, say 500 ohms or less. If this resistor is made too large, the input to the system may tend to drift to a high level even before the switch is closed. This is particularly true in a noisy environment. The value of resistor R1 may be obtained from the formula:

$$R1 = R2 \left[\frac{V_{IN}}{V_{OUT}} - 1 \right]$$

where V_{in} is the input voltage and V_o is the desired output voltage.

In this case, the input voltage is 12V, the desired output voltage is 5V and the value of R2 is 500 ohms. R1 will therefore be

$$R1 = 500 \ (12/5 - 1) = 700 \ \Omega$$

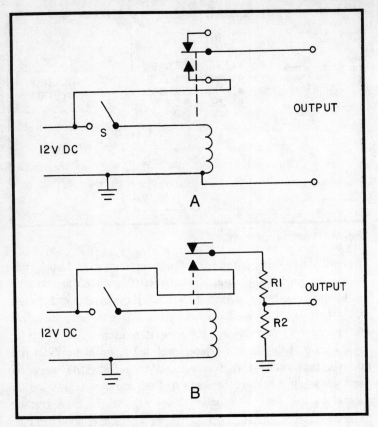

Fig. 14-2. Using a voltage to reduce voltage.

There are two limitations to this simple arrangement. One is that the sources might not be able to handle the current drain of the voltage divider. This current is simply

$$I = \frac{V_{IN}}{R1 + R2} = 10 \text{ mA}$$

The other limitation is that the circuit won't handle voltages that might go negative with respect to ground. Figure 14-3 shows a circuit that will provide an output voltage that swings between 0V and +5V when the input voltage swings between a negative and a positive voltage. The base resistor R1 is chosen to keep the base current of the transistor at a safe value when the input voltage is at its maximum positive value. The value of R1 is given by

$$R1 = \frac{V_{MAX}}{I_{b(MAX)}}$$

where $i_{b(max)}$ is the maximum permissible value of base current of the particular transistor used, and V_{max} is the maximum positive value that the input voltage can take.

This circuit operates because whenever the input voltage exceeds about 0.7V the transistor will turn on. Then the collector will be at very close to ground potential. When the input voltage falls below 0.7V the transistor will turn off. The collector will then be at the full +5V power supply potential. Therefore the output voltage will swing between 0V and +5V.

The circuit of Fig. 14-3 has two limitations. The first is that if the input voltage should swing more negative than about −6V the base emitter junction of the transistor will break down. This in itself isn't harmful, but if the transistor remains in this state very long, the resulting heat may cause damage to the transistor. Another limitation is that if the input voltage swings between two negative voltages, the transistor will never turn on and there will be no output.

The first of these limitations can be overcome simply by adding a diode between the base of the transistor and ground as shown in Fig. 14-4. Whenever the input voltage swings negative in this circuit, the diode will conduct protecting the transistor from the breakdown voltage.

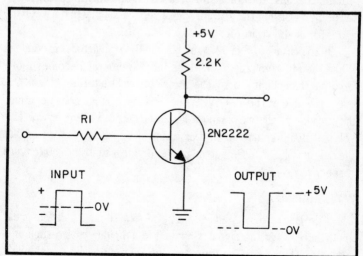

Fig. 14-3. Transistor level shifting circuit.

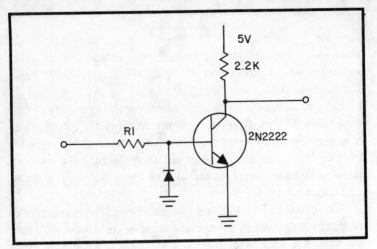

Fig. 14-4. Level shifting circuit with diode protection.

Note that both of the circuits of Figs.14-3 and 14-4 will invert the input signal. This isn't a serious problem. It can be corrected by installing an inverter in the line, or by an appropriate adjustment in the computer program.

Sometimes a sensor will provide a signal that never goes positive at all. It will swing between 0V and some negative value of voltage. The circuits of Figs.14-3 and 14-4 aren't much help in this situation because the transistor in the circuit would never turn on. A circuit that will shift the level of such a negative signal to between 0V and +5V is shown in Fig. 14-5.

In the circuit of Fig. 14-5 the base of the transistor is grounded. When the input voltage which is fed to the emitter becomes more negative than about −0.7V the transistor will be turned off and the output voltage will be at +5V. When the input voltage goes negative, the transistor will turn on and will pull the output voltage down. The voltage divider consisting of resistors R1 and R2 are chosen so that the output voltage will not go negative, or only slightly negative, with respect to ground.

WAVESHAPING

There are two properties of a pulse type signal that are of great importance when the signal is applied to a digital system such as a computer. These are the rise time and fall time of the signal which are identified in Fig. 14-6. Signals that are applied to a computer

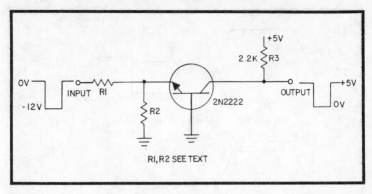

Fig. 14-5. Level shifter for negative voltage.

should have very short rise and fall times. Otherwise the computer might misinterpret the signal.

Figure 14-7 shows a logic gate representing the input of a computer and an input signal having comparatively long rise and fall times. When the input voltage is below 0.8V the gate interprets it as a logical low signal. When the input is above 2.4V, the gate interprets it as a logical high signal. When the input voltage is between these two values, the gate acts like an amplifier having a very high gain. If the input voltage stays in this no-man's land longer than the time required for the signal to propagate through the gate—about 10 nanoseconds—the gate will tend to oscillate. This is shown in Fig. 14-7.

The big disadvantage of oscillations on the input signals is that the computer might interpret each of the oscillations as a separate input signal and become thoroughly confused.

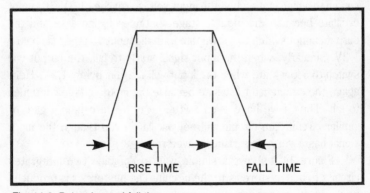

Fig. 14-6. Pulse rise and fall times.

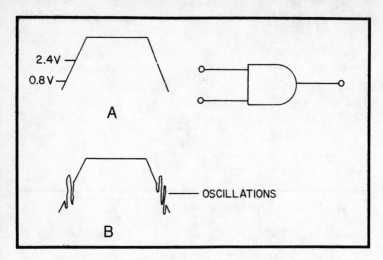

Fig. 14-7. Oscillations on leading and trailing edges of a pulse.

Inside a computer, all of the pulses have very short rise and fall times so there are no problems. In the real world of sensors, however, rise and fall times longer than 10 nanoseconds are common. Before these signals can be applied to a computer they must be shaped so that the rise and fall times will be within the proper limits.

One way to change a pulse waveform with slow rise and fall times to one that has short rise and fall times is to use a gate type of circuit that has some *hysteresis*. The behavior of the circuit is shown in Fig. 14-8. This circuit has an output that stays low until the input reaches some threshold level such as 1.3V. The output then goes high and stays high until the input voltage falls below some *lower voltage*, for example 1.2V. As the input voltage rises the output will switch to a high state when the input voltage reaches 1.3V. It cannot oscillate because the input voltage continues to increase and the output cannot switch to a low state until the input voltage falls below 1.2V. Similarly, when the input signal starts to fall, the output will switch to a low state when the input voltage falls below 1.2V. Here again, the circuit can't oscillate because the input voltage continues to fall. Thus if we have even a little bit of hysteresis in a gate or similar circuit, the circuit will not oscillate even though the input signal has a very slow rising waveform.

Figure 14-9 shows a simple way to combine two integrated circuits to add hysteresis to the circuit. The hysteresis is produced by the positive or regenerative feedback through resistor R2. When the input to the first inverter is low, its output will be high and the output of the second inverter will be low. Inasmuch as both ends of resistor R2 are at, or very nearly at, the same potential little or no current will flow through it. When the input to the first inverter starts to rise and rises to about 1.3V, its output will start to lower. This will make the output of the second inverter start to rise. Thus some current will flow through resistor R2 raising the input voltage even more. This whole process happens so fast that the output of the circuit jumps to the high state before either of the inverters has a chance to do any oscillating.

A circuit with the properties of that shown in Fig. 14-9 is called a Schmitt Trigger circuit. Integrated circuit gates are available that have Schmitt triggers built right into them. These circuits are ideal for conditioning slowly rising signals so that they will be suitable for using as computer inputs.

SWITCH DEBOUNCING

No practical switch or relay makes a really firm contact the moment that its contacts close. What happens is that when the

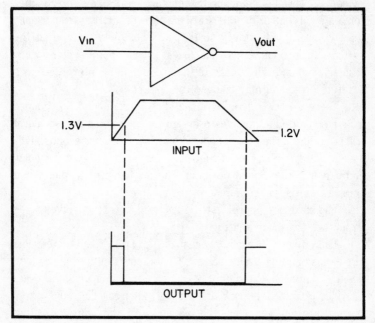

Fig. 14-8. Inverter with hysteresis.

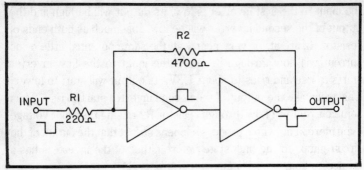

Fig. 14-9. Using positive feedback to get hysteresis.

contacts of a switch or relay come together, the moving contact bounces back thus opening the switch for a very short period of time. Most switches and relays will bounce several times before settling down to the closed position. In one control system a toggle switch would actually bounce over 200 times before settling in the closed position.

This bouncing of the contacts of a switch or relay is usually very rapid and the switch settles into its final position is a few milliseconds. This bounding of the contacts is not a problem in a conventional control system. All of the parts of such a system usually react comparatively slowly so they will not even notice that the switch actually bounces when it closes. In a computer system which reacts very fast each bounce of the switch will be interpreted as a separate pulse and may confuse the computer.

There are two general approaches to removing the bounce from a switch. One involves programming the computer so that it will not react to the state of a switch for about 5 milliseconds. This usually will give the switch contacts time enough to stop bouncing and settle in the closed position. The approach is simple, but it complicates the programming of the computer.

The hardware approach to debouncing a switch involves some circuit like that shown in Fig. 14-10.

This circuit is basically a latch made up of two NAND gates. The circuit has only two stable states. The output is either high or low. The coupling from the output of one gate to the input of the other provides positive feedback that makes the circuit flip rapidly from one state to another. When the switch is up in the figure the output will be in a low state. When the switch is moved down, the output of

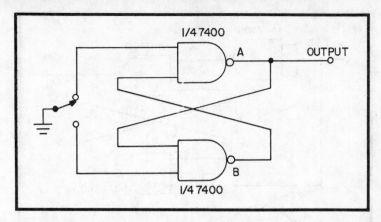

Fig. 14-10. Switch debouncing using two gates.

the bottom gate will start to a low state. The fact that this output is coupled to one of the inputs of the top gate tends to drive its output high. This in turn reinforces the action of the lower gate. The result is that when the switch is moved from one position to the other the output rapidly changes state. It then latches in its new state and even if the switch bounces it will have no effect on the output of the stage.

The principle limitation of the circuit of Fig. 14-10 is that it requires double-throw switches. Many sensing switches are more conveniently made as single throw switches. This limitation can be overcome by adding another gate to the circuit as shown in Fig. 14-11. Here the first gate simply acts as an inverter so that different

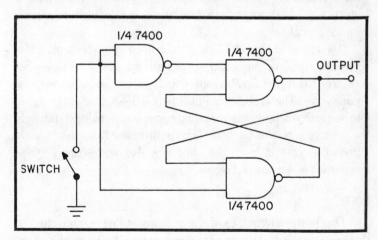

Fig. 14-11. Debouncing a SPST switch.

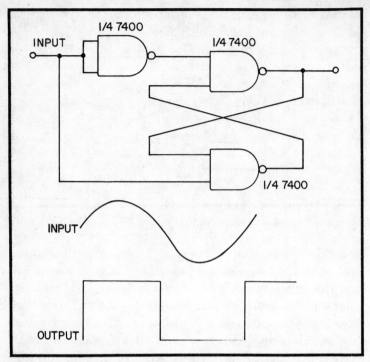

Fig. 14-12. Using the circuit of Fig. 14-11 for wave shaping.

polarity signals will be applied to the two inputs of the latch. There is no need to connect the switch to a logical high, because in TTL logic when the input of a gate isn't connected to anything, it will assume a high state. Thus all we have to do is to connect the input to ground with the switch.

The circuit of Fig. 14-11 can also be used for waveshaping. If a sinusoidal signal or any other signal that changes state slowly as compared to a digital signal is applied to the input, the output will be a square wave. This is shown in Fig. 14-12. The shaping of the signal into one with sharply rising and falling edges is accomplished through the positive action of the latch. The positive feedback in the latch shapes the wave in much the same way that waveshaping is accomplished in a Schmitt Trigger.

NOISE

One of the advantages of digital systems that is often cited is that because the digital signal has only two possible levels the effect of noise on the signal can be minimized. Although this is true, it may lull the unsuspecting technician into believing that digital systems are not susceptible to the influence of noise. This isn't true. Although a digital signal, once it gets into a system, may not be corrupted by noise the way it would in an analog system, noise that does get into a digital system is often interpreted by the system as a signal.

When a digital component such as a microprocessor is connected to other digital components on a printed circuit board, noise isn't usually a problem. Where the trouble comes in is when we make connections to the outside world. These external connections often couple noise and power line hum into the system and the system thinks that the noise is a signal. The result is usually erratic and completely unpredictable operation. One of the major considerations in the design of an interface is to eliminate the effect of noise.

There are many ways in which noise can get into a computer system. The one that concerns us is noise entering through our connections to the outside world, such as connections to sensors and control devices. One insidious problem is shown in Fig. 14-13. This situation is commonly known as a ground loop.

This circuit could represent connections to a sensor mounted on something that is motor controlled. Suppose that the motor causes some sort of signal such as the 60 Hz voltage or commutator noise to flow through the ground. On the surface, this seems harmless, but a closer inspection of the circuit shows that the common lead between the computer and the sensor is actually in parallel with

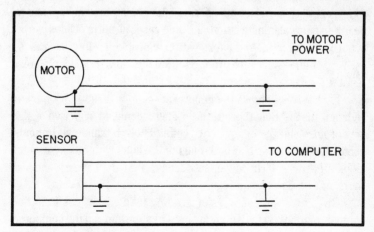

Fig. 14-13. Ground loop in sensor leads.

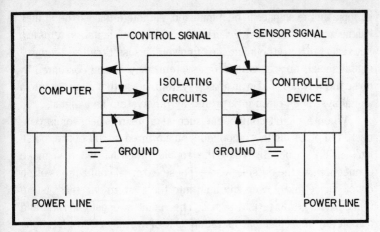

Fig. 14-14. The principle of isolation.

the ground. Thus any current in the ground will divide with some of it passing through the common lead to the sensor. This lead will have some internal impedance, even though it might be small, and there will be a voltage drop across it. Thus we have managed to introduce the noise or hum voltage right in series with the signal from the sensor.

One way to avoid ground loops is to have ground connections only at one of the leads. Unfortunately this isn't always possible.

The best way to keep noise and extraneous signals out of a computer is to use some form of isolation between the lines from the sensors and the computer. The type and amount of isolation depends to a great extent on just how serious the noise problem happens to be in any particular installation. In some cases all sorts of lines can be run from sensors to computers with few problems. In other cases, the sensor lines will cause problems until they have been completely isolated.

Figure 14-14 shows the general concept of isolation. Complete isolation means that there is no metallic connection between the sensor lines and the computer circuits. Thus we can have grounds connected to the earth at both ends of the lines without any ground loops. Separate 60 Hz power lines can be connected at each end and there will be no induced hum.

One of the easiest ways of isolating the sensor lines from the computer circuits is to use a relay between the line and the computer as shown in Fig. 14-15. The power for operating the relay coil comes

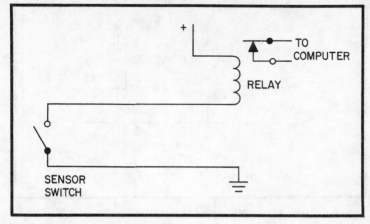

Fig. 14-15. Using a relay for isolation.

from a power supply that is built into the interface. The only connection to the computer circuits is through the relay contacts.

Another useful method of coupling between circuits with complete isolation is by means of an optocoupler. The optocoupler, shown in Fig. 14-16, consists of a light emitting diode and a phototransistor mounted in the same case. When there is no input current, the phototransistor will be dark and no collector current will flow. The output will then be at the +5V collector voltage. When input currents flows, the LED will illuminate the phototransistor. This will

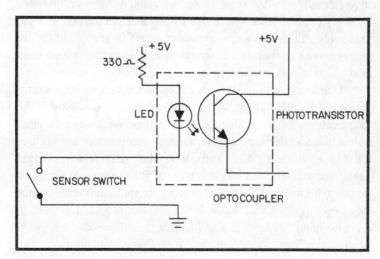

Fig. 14-16. Isolation with an optocoupler.

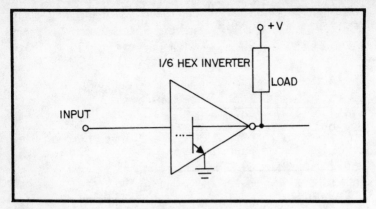

Fig. 14-17. Using a buffer to increase load current.

turn the phototransistor on bringing the output to almost ground potential.

GETTING CONTROL SIGNALS OUT OF THE COMPUTER

The problem of getting the control signal that is to control something out of the computer into the real world is somewhat similar to the problem we just discussed of getting signals from the sensors into the computer. We still have the possibility that noise on the output lines will get back into the computer. We don't usually have to worry about the waveshape of the signal from the computer. It will usually vary between 0V and +5V and most control devices are very tolerant of waveform. We usually find that the output of the computer, which will only drive about one TTL gate input, is not heavy enough to operate a control device. This means that we need some sort of buffer to increase the signal.

One handy form of buffer is the open-collector inverter shown in Fig. 14-17. This integrated circuit has only one transistor in the output instead of the usual totem pole arrangement. When the input is low, the output transistor is turned off and the output line will take on the value of the supply voltage. When the input is high, the output transistor will be turned fully on and it can sink a great deal of current to ground. The Type 7405 Hex inverter can safely sink about 16 mA of current to ground from a +5V supply. This is enough current to operate many homemade latching relays. If more current is required, the Type 7416 Inverting Hex Driver can be used. The connections are the same as in Fig. 14-17, but the 7416 can safely sink up to about 40 mA and can operate from an output power supply of up to +5V. This is enough current to drive almost any relay that will be used in a control system.

When still more current is required, an inverter can be used to drive a transistor as shown in Fig. 14-18. Here we can get any amount of current depending on the size of transistor that we use. Of course, if a large power transistor is used, it will be necessary to use the proper heat sink. Resistor R1 in the figure is selected to provide enough base current to turn the transistor fully on. When the output of the inverter is low, the current that normally flows into the base of the transistor flows through the output transistor of the inverter.

With the arrangements of Figs. 14-17 and 14-18, the small signal from the computer can be made to drive almost any of the control devices that we have described in earlier chapters. Each of these circuits inverts the signal from the computer, but this is no problem because it can easily be handled in the programming of the computer.

INTERFACE POWER SUPPLY CONSIDERATIONS

In the control systems described in the earlier chapters of this book the power supply hadn't been particularly critical. In the interface where TTL integrated circuits might be used both the power supply and the power supply lines can indeed be critical and require special attention. The first consideration is that the power supply lines must be liberally bypassed. The best approach is to use a 0.01 disk ceramic capacitor from the positive line to the ground connec-

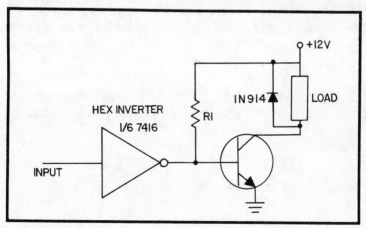

Fig. 14-18. Driving a transistor from the computer.

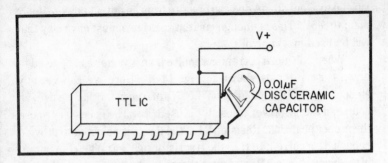

Fig. 14-19. Bypassing digital ICs.

tion of the integrated circuit as shown in Fig. 14-19. Ideally, there should be a separate capacitor at each integrated circuit. Usually these capacitors are called "despiking" capacitors. Their purpose is to minimize power supply transients that might occur when the circuit switches from one level to another.

These transients occur simply because the TTL integrated circuits switch so rapidly. Bypassing back at the power supply usually doesn't do any good at all because the inductance of the leads presents a high reactance at the frequency of the switching waveform.

The power supply used for the TTL circuits in the interface must be well regulated and must be capable of supplying the necessary current. Figure 14-20 shows the diagram of a suitable supply. The regulation is provided by the integrated circuit voltage regulator. The input filter capacitor should be large enough to handle the necessary filtering. The output capacitor is a little more critical. It helps to keep the regulator stable. A good grade tantalum capacitor should be used. Although not indicated on the schematic, the voltage regulator will require a heat sink if more than a small current is drawn from the supply.

The Type 7805 voltage regulator also has internal current limiting so it will not be damaged by accidental short circuits.

SUMMARY

Interfacing is probably the most important consideration when using a computer or microprocessor in a control system. Unless proper attention is given to the considerations discussed in this chapter, the system will probably not work reliably, and in some instances, the microprocessor may be damaged.

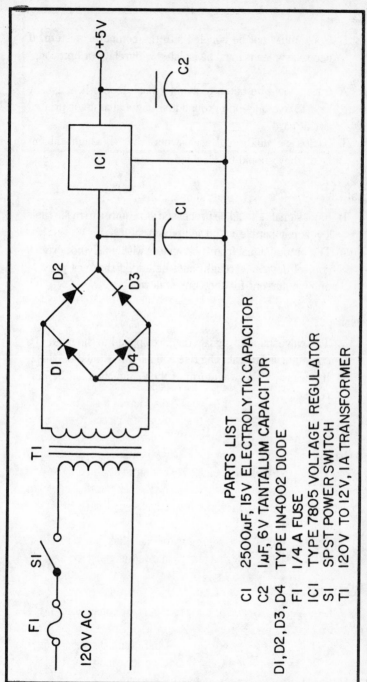

PARTS LIST

C1   2500μF, 15V ELECTROLYTIC CAPACITOR
C2   1μF, 6V TANTALUM CAPACITOR
D1,D2,D3,D4   TYPE 1N4002 DIODE
F1   1/4 A FUSE
IC1   TYPE 7805 VOLTAGE REGULATOR
S1   SPST POWER SWITCH
T1   120V TO 12V, 1A TRANSFORMER

Fig. 14-20. Power supply for interface circuits.

269

Noise

1. Noise must not be coupled into the computer system. If necessary, leads must be isolated with relays or optocouplers.
2. Care should be taken to avoid ground loops. It is usually best to use only one ground point on a chassis or printed circuit card.
3. If lines are run through areas where there is a high ambient field, they should be shielded.

Signal Levels

1. Input signals should be processed as required to make their levels compatible with the computer.
2. The output signal from a computer will usually not drive a control device. Circuits must be added that will provide enough current for the control device.

Waveshapes

1. The waveshape of signals that are applied to the input of a computer is critical. The rise and fall times must be short. If necessary the waveform of these signals must be corrected.

15

Finding the Components

We have pointed out several times that few if any home control problems are exactly alike. Because of this there are always a few required components that never seem to be available. These hard to find components are usually mechanical linkages, hydraulic components, and special electrical components. Even when a commercial source is finally located for the hard to find component, the price is often beyond the reach of the average experimenter.

The answer to this problem is that usually the needed component is available locally at a reasonable price if it can be salvaged from some other application. The average electronics experimenter is often adept at scrounging electronic components but he just doesn't know where to look for mechanical, electrical, or hydraulic components. Often he doesn't know exactly what to look for. All he knows is that he has some sticky problem such as automatically opening a window, and that he needs something that will handle the job, but he doesn't know exactly what it will be. In this chapter we will point out the different things that can be salvaged from a junkyard or second-hand store.

THE SURPLUS STORE

Most cities of reasonable size have what is called a salvage store or a surplus store. Many of these originated as outlets for

Table 15-1. Sources of Industrial Components.

HERBACK & RADEMAN, INC. (MIN. ORDER $10)

401 EAST ERIE AVENUE

PHILADELPHIA, PA 19134

AMERICAN DESIGN COMPONENTS (MIN. ORDER $15)

39 LISPENARD ST.

NEW YORK, NY 10013

C AND H SALES CO (MIN. ORDER $15)

2176 E. COLORADO BLVD

PASADENA, CA 91107

government surplus equipment that was dumped on the market after World War II and the Korean conflict. As the government surplus material became scarce, many of these shops managed to stay in business by dealing in second hand industrial equipment. Even if such a store doesn't have a particular component, the operator may well know of a source. These stores can be found listed in the yellow pages of the telephone directory.

Table 15-1 lists the names and addresses of a few companies that deal in industrial components that may be either new or surplus. These houses occasionally print fliers and catalogs of items that they offer for sale. They are often the source of just the thing needed to complete a control system. Occasionally they have a component that almost meets a need. Sometimes such a component can be modified by a local machine shop to meet the need exactly.

HOUSEHOLD APPLIANCES

Household appliances are becoming more complicated each year as new features are added to increase their convenience value. Many of these added features amount to automating something that was formerly done manually. For example, at one time every re-frigerator had to be defrosted manually by disconnecting the power and allowing the unit to warm up enough to melt the accumulated ice.

Most refrigerators manufactured in the past few years will defrost automatically. This means that something has been added that will handle the defrosting function automatically. In many cases this something that has been added is a 24-hour timer that can rather easily be converted to timing something that is controlled by the home control system.

This same philosophy should be applied to all appliances. The experimenter should give thought as to how each appliance performs each of its functions and to whether or not any of the equipment can be salvaged for use in a control system. Appliances are frequently junked when many of their features are still functional. Second-hand appliance dealers usually have a wide variety of such appliances and are happy to find a market for them or their components at a reasonable price.

The same method of thinking should be applied to automobiles. Whenever some function in an automobile is performed automatically or remotely, the experimenter should ask himself if the components that perform this function could be adapted to perform a similar function in a remote or automatic control system. For example, in some automobiles the radio antenna is raised and lowered by simply pressing a switch. The question is can the mechanism that performs this function be used to raise and lower something in a control system. The answer is yes. The electric motor used to raise and lower a radio antenna in an automobile can rather easily be adapted to opening and closing home draperies by remote control.

In the remainder of this chapter we will review some of the components that might be found in home appliances and automobiles that can often be used to solve control problems. The list is by no means complete. The experimenter with a little ingenuity can readily think of many other devices that have components that can be adapted for use in control systems.

THE REFRIGERATOR

At first glance it would seem that the household refrigerator with its sealed unit would have nothing to offer an experimenter. Let's take a more detailed look. Many refrigerators with automatic defrosting have a 12- or 24-hour timer that turns off the cooling unit at some specified time so that the unit can defrost. This same timer may still be good in a refrigerator that has been discarded for some other reason. If so, it might well be used for timing something that is handled by a control system. Other automatic defrosting refrigerators have a counter that counts the number of times the door of the refrigerator is opened and opens a switch after the door has been opened a predetermined number of times. This too might be used to advantage in a control system.

Every refrigerator has a door switch that closes when the door is opened to turn on the inside light. A switch of this type can be used as a sensor to tell when a door or window or anything else that moves is open or closed.

Many refrigerators are equipped to make ice cubes automatically. The mechanism that performs this function includes a solenoid valve that controls the flow of water into the ice cube making chamber. This valve can be used to control the flow of water anywhere.

All refrigerators have some sort of thermostat. Often a thermostat of this type can be used to detect freezing conditions. For example, it might be desirable to keep a storage room unheated except for times when the temperature might fall below freezing. A thermostat that closed when the temperature approached freezing could well be used to control the temperature in such an area.

WASHING MACHINES AND AUTOMATIC WASHERS

Automatic washers have many components that can be used for other purposes. The most obvious is the motor which is usually a fractional horsepower AC type. This motor will rotate too fast to drive most things directly, but it can be used as a source of mechanical energy to drive something that requires quite a bit of power. The pump that is used to drain water from the washer might be used in connection with the motor to pump water from a leaking basement. The pump is usually a centrifugal type that will pump water at a reasonably fast rate. The pressure switches that detect the water level in a washer might also be used in connection with the motor and pump to make up a completely automatic sump pump that will turn on when the water in a sump in the basement reaches a certain level.

An automatic washer has solenoid valves that control the flow of hot and cold water into the machine. It often also has a few mechanical solenoids that can be used where something is to be moved over a short distance. Often the machine includes one or more relays that can handle quite a bit of current. The timer used to control the washing cycle can sometimes be used for other purposes. Timers of the type used in automatic washers have a motor that can be detached. If a similar but slower motor can be installed the timer will operate over a much longer timing cycle. Many washers have attractive switches on the front panel that would dress up the control panel of a remote control system. Top loading washers often have mercury switches that will interrupt power when the cover is opened.

CLOTHES DRIERS

Like the automatic washer, the clothes drier has a motor that might be used for some other purpose. It also has a timer and sometimes has attractive switches on the panel. In addition it often has a door switch that controls an internal light. This switch can be used to sense the position of something.

Most driers have a limit thermostat that will shut off the heating element when the internal temperature exceeds a limit. Depending on its construction, it may be possible to set this thermostat to operate at a different temperature. Usually there is a relay that will handle a high current to control the heating element. Even the blower might be used for some purpose.

KITCHEN RANGE

A discarded kitchen range will often yield many components that can be used in a control system. Often relays are used to control the current to the various heating elements. These relays will handle many amperes of current. Of course, the timer on a range is usually a 24-hour timer and has at least one on-off set of contacts. Often there is one setting for cooking time and another for the time that cooking will start. Such a timer can be used to time the operation of almost anything. The range timers salvaged from a junkyard or second-hand store are often very inexpensive and can often be reconditioned so that they are nearly as good as new.

Many ranges have circuit breakers that will open the circuit if the load current exceeds a certain value. Such circuit breakers provide good protection of any circuit where the source is capable of providing enough current to constitute a fire hazard.

AUTOMATIC DISHWASHER

Even the automatic dishwasher may be a source of useful components. All dishwashers have pumps that recirculate water and drain the unit. These might be used for automatic sump pumps. The dishwasher also contains a timer that might be useful for some purpose. Many dishwashers contain relays that can be used to advantage.

Another device that can be salvaged from a dishwasher is the switch that shuts off the water when it reaches a certain level. This is often a pressure operated switch that can be salvaged. It can usually be used as some sort of sensing switch.

Every dishwasher contains solenoid operated valves that control the flow of water into the machine. These can also be used to control the flow of water in a control system.

AUTOMOBILE

There are probably more automatic and remote controlled systems on a luxury automobile than on any other one device. Figure 15-1 shows a sketch of an automobile with many of the possibly useful components called out. Of course, there are many differences between types of cars and a device that can be salvaged from one make may be completely different from the device that performs the same function in another model. Thus, one should become familiar with how the subsystems of all different makes of cars work. What may be vacuum controlled in one car might be electrically controlled in another. The operator of an automobile junkyard can be a very useful source of this type of information.

Probably the most obviously useful component in an automobile is the switch. There are many different types of electric switches used in cars. The dashboard usually has attractive switches that can be used as control switches. In many cars there are rocker-type single-pole double-throw switches that can be used with a low voltage control system. There are plunger operated switches that turn on lights when doors, trunk or hood are opened. These switches can be used as sensing switches to detect the position of anything that moves. The brake light switch is a pressure operated switch that can be used to detect the pressure in a fluid. It might make a good indicator switch for a hydraulic system.

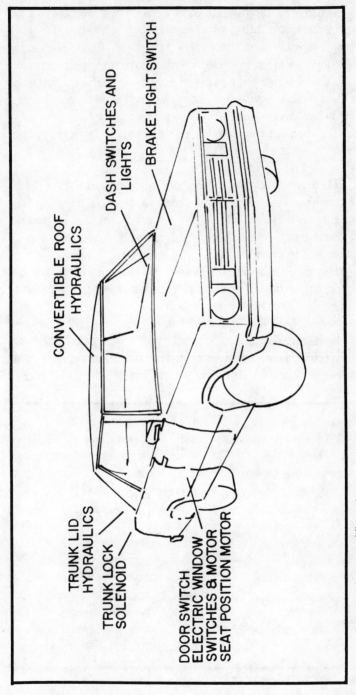

CONVERTIBLE ROOF HYDRAULICS

DASH SWITCHES AND LIGHTS

BRAKE LIGHT SWITCH

TRUNK LID HYDRAULICS

TRUNK LOCK SOLENOID

DOOR SWITCH
ELECTRIC WINDOW SWITCHES & MOTOR
SEAT POSITION MOTOR

Fig. 15-1. Salvageable automobile parts.

Switches that are used to control electric windows are usually single-pole double-throw switches that can be used with a low voltage control system. Usually they are quite attractive. Some cars have a joy-stick type of switch that is used to control the operation of an electric motor that moves the seats of the car. This switch could well find application where a control system is used to move something in two directions.

The dash of many cars has many indicator lights that can be used with switch type sensors. These too are usually attractive and can be used on a control panel.

There are usually several electric motors in a car. Usually these are in reasonably good condition when a car is junked because of an accident. Such motors include the electric motor that is used to position automatic seats. Usually this motor has a high torque and can be used to provide a strong pull or push.

The drive mechanism that is used with the seat motor can often be modified for other applications. It is often easier to make minor changes in such a mechanism than to try to build one.

Figure 15-2 shows a sketch of an electric motor used in many cars to raise and lower the radio antenna. This unit usually consists of an electric motor that drives a nylon rod which, in turn, raises and lowers the antenna. This arrangement can be used with minor

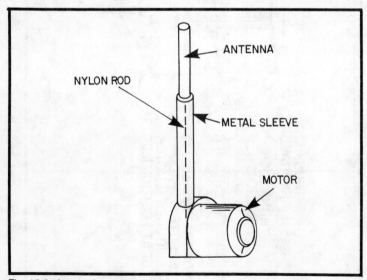

Fig. 15-2. Automatic antenna motor.

modifications to handle things such as draperies. Usually the arrangement will provide a pulling force of ten to fifteen pounds.

The highest power electric motor in a car is the starting motor. This motor is rated for intermittent duty and couldn't be used to drive anything continuously, but it is available for those applications where a lot of power is required for a short period of time. There probably aren't many control applications that require this much power, but it could possibly be used in a homemade garage door opener.

The motors that are used to drive electrically operated windows are also useful in control applications. One nice feature of these motors is that they are used with a mechanism to raise and lower windows. Sometimes the mechanism can be modified for another application. As in the case mentioned earlier, it is often easier to adapt an existing mechanism for another purpose than it is to attempt to build a new mechanism for a particular application.

There are other automotive electric motors used for various purposes. Sometimes the motor and blower used for rear window defrosting can be used without change for some application.

Almost every car has a starting relay that controls the current to a starting motor. Again, most control systems will not need a relay having this much current carrying capacity, but if for any reason a starting motor is used in a system, the starting relay can be used to control it. Often there are other smaller relays that can be used to advantage.

Closely related to relays are solenoids. These are not as widely used as some of the other electric components, but some cars have solenoids that lock the trunk. There can be adapted for use as remotely controlled door locks. Sometimes a solenoid is used to advance the throttle whenever the air conditioner is turned on. With a little ingenuity a special solenoid might be made from a starting relay. Such a solenoid would not provide a very great length of travel, but would exert considerable force. Inasmuch as the relay is not rated for continuous duty, it should not be used to hold anything in position for an extended period of time unless the current is reduced.

The solenoids that are used to lock seat backs in two door cars are interesting. This is a high current solenoid that is used to actually unlock the seat backs. This solenoid draws ten amps or more and

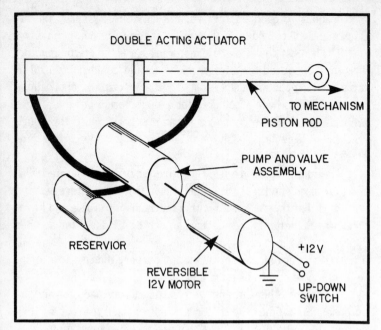

Fig. 15-3. Hydraulic lifter for convertible roof.

provides a strong force. Once the seat back is unlocked, the heavy current coil is de-energized and a smaller control that draws less than an ampere of current is used to hold the solenoid in position. Thus this solenoid can be used where it is necessary to hold something in position for an extended period of time.

The solenoids used in electric door locks are very useful in that there are two separate windings. One locks the door and the other unlocks it. No current is needed when the lock is either locked or unlocked. The only time that current is drawn is when the state of the lock is being changed. Thus this type of solenoid can be used where something is to be in one or the other position for a long period of time. No current is required except when it is being changed.

Figure 15-3 shows a diagram of the hydraulic system used in some convertibles to raise and lower the roof. The system consists of a motor driven pump, a fluid control valve and two double acting hydraulic actuators. This system is capable of producing forces of between 300 and 400 pounds per square inch and can be used wherever a large force is required. The output is a linear motion which makes it much easier to apply than a motor where some means would be required to convert the rotary output of the motor to a linear motion.

In operation, the motor turns one direction to raise the roof and in the other direction to lower it. When the switch is in the "up" position, the pump rotates in such a direction as to inverse the pressure at the bottom of the cylinder force forcing the piston to move forward. The fluid at the top of the cylinder is allowed to flow back into the reservoir by a control valve that is located inside the pump housing. This action forces the piston out of the cylinder with a high force.

When the switch is in the "down" position, the motor rotates in the opposite direction. The pump then applies hydraulic pressure to the top of the cylinder. This forces the piston back into the actual actuator. The fluid at the bottom of the cylinder is allowed to flow back into the reservoir by the internal control valve.

Figure 15-4 shows a sketch of the actuator. It is mounted at the top by a bolt which passes through the end of the actuator. In this way, it is allowed to swing back or forth if required while the piston is moving in an out. This arrangement simplifies the mechanical requirements.

Figure 15-5 shows such an actuator can be mounted for opening and closing a door. The actuator is mounted near the hinge side of the door. As the piston moves out of the actuator it will force the door open. As the door opens, the actuator will swing so that the piston rod can follow the motion of the door. The mechanical arrangement must be such that the mechanism will not bind, because the actuator can produce enough force to break the door if the motion should be impeded.

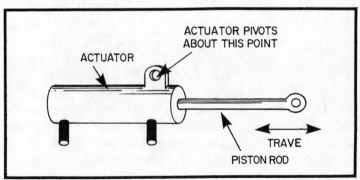

Fig. 15-4. Detail of actuator of Fig. 15-3.

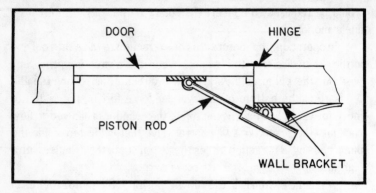

Fig. 15-5. Actuator used as a door opener.

With a hydraulic door opener, the door must be unlocked before the pump motor is energized. This can be accomplished by using a solenoid door lock that must be energized before the motor can be energized.

POWER SUPPLIES FOR OPERATING AUTOMOTIVE COMPONENTS

As we have seen in the preceding paragraphs, the automobile is a good source of components for control systems. These components will almost always require a 12V DC power supply. Because of the fact that many automotive components deliver a lot of mechanical power, the currents are often high compared with the currents drawn by other electronic components. It isn't at all unusual to find motor and solenoid currents of 10 to 20 amperes for short periods of time. For this reason the power supply should include a storage battery.

There are advantages and disadvantages to using a power supply with a storage battery. The big advantage is that the system will operate even during an electrical power outage. This is particularly useful when the power outage is caused by some emergency such as a fire. Another advantage of using a storage battery is that the power supply can deliver very large currents for short periods of time while not drawing very much current from the power line.

The disadvantage of using a storage battery in a power supply is that is requires special handling and maintenance. An ordinary lead acid storage battery should be located where there will be no serious damage from small spillage of the electrolyte and so that there is no fire or explosion hazard from the hydrogen gas that is released during the charging cycle. Most of these disadvantages can be

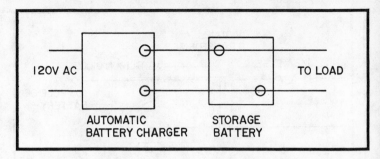

Fig. 15-6. Power supply for automotive components.

overcome by using one of the new sealed storage batteries that are being sold for use in automobiles.

The simplest power supply that can be used for powering automotive components is shown in Fig. 15-6. It consists merely of a storage battery and an automatic charger that will cut off when the battery is fully charged. Both items are commercially available. The limitation of this arrangement is that the charger may cause interference that will degrade the performance of the rest of the control system. This can usually be overcome with an interference filter of the type shown in Fig. 15-7.

Probably the easiest approach to keeping a storage battery charged is to use the trickle charger shown in Fig. 15-8. This circuit consists only of a 12-volt transformer and a rectifier diode. The circuit will provide some charging current because the peak value of the secondary voltage is about 17.8V. When the battery is fully charged, the charging current will be one ampere or less. As the

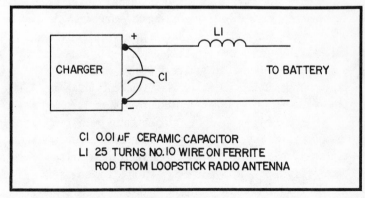

CI 0.01 μF CERAMIC CAPACITOR
LI 25 TURNS NO.10 WIRE ON FERRITE
   ROD FROM LOOPSTICK RADIO ANTENNA

Fig. 15-7. Interference filter for battery charger.

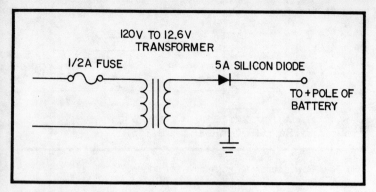

Fig. 15-8. Simple trickle charger for storage battery.

battery discharges, the current will increase to two or three amperes.

Even though the drain of the control system on the storage battery can be very high for short periods of time, the control system will be used intermittently so the amount of discharge will be small. The charging current from the trickle charger will be small, but continuous. If the battery becomes partially discharged during the day, the trickle charger will catch up with it at night.

16
Radio Control Systems

This is intentionally the last chapter of the book. While it is recognized that there are applications where a radio link is by far the best answer to a control problem, the number of both technical and legal problems that arise in connection with these systems often make it more attractive to take another approach.

LEGAL CONSIDERATIONS

The transmitter of a radio control system, even though it operates at a low power level, is actually a radio transmitter and is subject to regulation by the Federal Communications Commission. Many people mistakenly think that if the power level is low enough, the control system is not considered to be a radio transmitter. This simply isn't true. Any device that radiates signals as a part of its function is a radio transmitter regardless of its power level. In general radio transmitters must be licensed. Exceptions to the licensing requirement are contained in Part 15 of the FCC rules. Anyone considering building a radio control system should be thoroughly familiar with this part of the FCC rules.

Many commercial transmitters of the type used for radio control of devices such as garage door openers are Type Approved, Type Accepted, or Certified by the FCC and can be used without a license. It should be noted, however, that it is illegal to make any

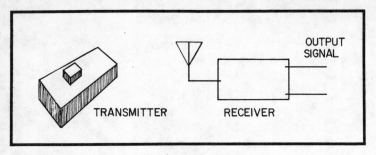

Fig. 16-1. Radio controlled garage door opener.

modifications to such a device. A frequent violation involves connecting a longer antenna to a type approved device. This voids the type approval and makes the device illegal.

In order for the transmitter of a radio control system to be legal it must meet the requirements of Part 15 of the FCC rules. These requirements include the frequency of operation, the level of radiation, and the amount of harmonic or spurious radiation. Usually the home constructor doesn't have the equipment to assure that these conditions are met.

With the increasing amount of policing activity by the FCC in connection with the Citizen's Band, it is quite likely that an FCC monitor will show up in almost any location.

INTERFERENCE PROBLEMS

Any radio transmitter has the potential for causing interference to radio, television, or commercial radio services. Such interference is illegal and may result in legal problems as well as in deteriorating relationships in the neighborhood.

Equally serious is the fact that any radio receiver, including the receiver of a radio control system, is subject to interference from other radio signals. This can make an otherwise excellent control system nearly worthless. It is easy to imagine the chaos that would result if the various things connected to a control system started to operate erratically during the middle of the night when a CB operator turned on an illegal linear amplifier. Before a radio control system can be trusted, it should be tested to be sure that it neither causes nor is susceptible to interference. Again, the home constructor is not likely to have the necessary equipment to do either of these things properly.

COMMERCIALLY AVAILABLE COMPONENTS

One solution to the problem is to use components that are commercially available. The most common radio control system is the electronic garage door opener. The transmitter of such a system complies with Part 15 of the FCC rules and is usually safe to use. Usually the system is quite free from interference effects. It neither causes interference, nor is overly susceptible to it.

Figure 16-1 shows a block diagram of a garage door opening system. As it is, without any addition of modification, it can turn a single device on or off. Usually this isn't enough flexibility for a home control system. Additional functions can be added without modifying the transmitter by adding selective circuitry to the output of the receiver.

Figure 16-2 shows how something like a garage door opener which is designed to control only one device can be used to control several devices. The output of the receiver, which is usually a simple contact closure, is fed to a counter. The outputs of the counter are decoded as described in Chapter 10 to provide a separate output for each control function. The outputs of the decoder are fed to one input of a NAND gate. The other inputs of the NAND gates are connected to the output of a timer.

In operation, the transmitter is keyed a number of times corresponding to the number of the device which is to be controlled. For example, if we want to turn on device number 4, we would key the transmitter 4 times. The counter then counts up to 4 and the

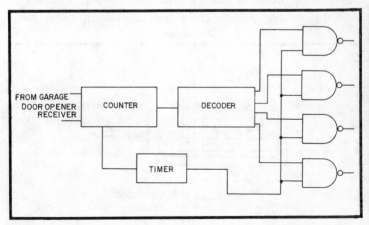

Fig. 16-2. Controlling more than device with a garage door opener.

decoder provides a high output on line 4. Nothing has happened so far in the control system, because the output of the decoder is connected to one of the inputs of a NAND gate. However at the time that the transmitter was first keyed a timer circuit starts. After a long enough time for the proper code to be keyed in, the timer will provide a positive pulse to all of the NAND gates. Only gate four will change state, however, because this is the only one driven by the decoder.

This circuit is quite simple. It may, however, be subject to switch bouncing and the output of the receiver may have to be debounced as explained in Chapter 14. The principle limitation of the circuit is that once the transmitter is keyed the control cycle is started and must be followed through. If the transmitter is keyed one or more times accidentally some control device will be triggered.

This limitation can be overcome by the ingenious experimenter by requiring two successive transmissions for the system to do anything. If only one set of pulses is transmitted, the circuit will automatically reset without changing any of the controlled devices.

Another approach to avoiding the legal requirements of a radio transmitter is to use a "wireless microphone" of the type designed to be picked up on the FM broadcast band. These devices have very low power and a very limited range. This is overcome to some extent by that the fact that standard FM receivers are available at a very low cost. It is economically practical to have more than one receiver in the house if necessary to get the desired range.

The transmitter of this system is shown in Fig. 16-3. Here a standard two-tone encoder of the type described in Chapter 7 is

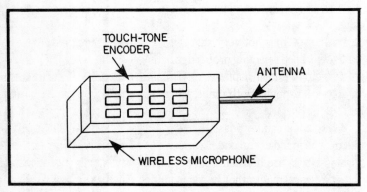

Fig. 16-3. Radio control transmitter.

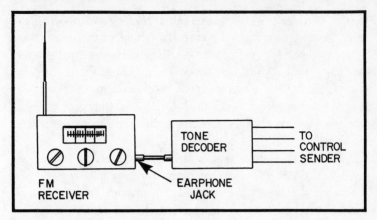

Fig. 16-4. Radio control receiver.

connected to a miniature speaker and the whole thing is mounted right on the wireless microphone. The transmitter is still legal because no connections have been made to it and nothing has been changed internally. The signal from the wireless microphone will now be an FM signal with the standard two-tone control signals as modulation.

The receiver, shown in Fig. 16-4 is a regular FM broadcast receiver with provision for using a headphone. The headphone output is connected to a tone decoder and the rest of the system is just like a tone operated system of the type described in Chapter 7.

This system is very easy to get operating and there is no problem with the legal requirements. Naturally the transmitter should be checked to be sure that it does not cause interference. This is unlikely, because the power is less than 100 milliwatts and the transmitter is only turned on when it is necessary to control something.

The system is not very susceptible to interference because of the interference rejecting properties of the two tone system. Even if an interfering signal happens to simulate one of the two tones, nothing will happen. Both tones are necessary to operate the system.

This book cannot possibly cover all the devices and controls that may be found useful under various circumstances. However the building blocks are here and the possibility of adapting these ideas and systems to meet other needs is limited only by the imagination and ingenuity of the technician or experimenter.

Index